JCER

Journal of Consciousness Exploration & Research

Volume 4 Issue 2
March 2013

Various Contents of Consciousness & Theories of Their Origins

Editors:

Huping Hu, Ph.D., J.D.

Maoxin Wu, M.D., Ph.D.

Advisory Board

ISSN: 2153-8212

Journal of Consciousness Exploration & Research
Published by QuantumDream, Inc.

www.JCER.com

Table of Contents

Article

The Creation of Happiness

Steven E. Kaufman*

ABSTRACT

This is not a world of suffering. To the contrary, it is a world where it is just as easy to create the opposite of suffering as it is to create suffering. One just needs to know which button to push on the machine of experiential creation. However, because most people are unaware of both how experience is created, as well as what it is they actually want, which is just to feel better, to create and apprehend a more wanted emotional experience, most people go about trying to create experience in a way that is the opposite of the way it is actually created, leaving them pushing the button of resistance rather the button of allowing, in which case they end up producing experiential results that are the opposite of those they intended to produce.

Key Words: pursuit of happiness, unhappiness, Existence, emotional experience, Vedantist, maya, Nature.

Happiness is not to be pursued, but must be created by the Individual who wants it. One can try and pursue happiness, and in that pursuit one may on occasion find it, but as often as not, or perhaps more often, one finds instead the opposite, i.e., unhappiness. The reason this is so is because in the pursuit of happiness one often uses resistance to try and get to where they think happiness is to be found, and in that resistance one creates unhappiness or unwantedness.

That is, when happiness or wantedness is pursued, as if it were something that exists independent and apart from the Individual that apprehends it, then the Individual sees happiness as being somehow inherent in some object or person or status, and they then pursue and go after the shiny object, thinking that once it is possessed that the happiness they believe to be somehow inherent in the thing will then be theirs. And once the person gets the thing, they may, for a moment, feel some happiness, but whatever happiness they feel is not actually coming from the thing they now seem to possess, no matter how much this seems to be the case. Rather, whatever happiness they feel they are themself creating as a result of their attitude of allowing with respect to the thing they now possess, and it is that attitude of allowing that creates happiness or wantedness, just as it is the attitude of resistance that creates unhappiness or unwantedness.

All experiential wantedness and unwantedness is created by the Individual that is apprehending that wantedness or unwantedness, according to a relation in which they are involved with their Inner Self, with their Soul, with their More Fundamental Individuality, with God, with their Source, with the Tao, with Absolute Existence, with whatever name one wants to give to that Aspect of Ourselves from which our Individual Consciousness flows and is projected, and from which Aspect our Individual Consciousness continues to flow and continues to be projected, as

*Correspondence: Steven E. Kaufman, Independent Researcher. http://www.unifiedreality.com
E-mail: skaufman@unifiedreality.com

long as we remain organically alive and thereby involved in the relations with Existence that create what we apprehend as physical reality.

The True Nature of the Individual is that the Individual, any Individual, is one pole of an Indivisible Duality of Existence, with the other pole of that Indivisible Duality being the Individual Existence that is projecting Itself, as our Individuality, from more fundamental levels of Existential Self-relation into the level of Existential Self-relation where the reality that we call physical reality or physical experience is created and apprehended. Thus, all Individuals are really two Individuals simultaneously; the Individuality we know as our self, and a *More Fundamental Individuality* that we either are completely unaware of, or that we think of as other than our self, as somehow separate from our self, as being of a different nature than our self, as being better than or superior to our self, when in actuality these two poles of Individuality are indivisible aspects of the same Existence, the same Realty. Put another way, all Individuals are really two Individuals, and those two Individuals are not two separate things, rather they are two poles or aspects of a single indivisible Reality.

These two poles of Existence, i.e., the Individual and the More Fundamental Individuality, correspond to the Atman and Brahman, respectively, and the indivisibility and identity of these two poles of Existence is recognized in the Vedantist philosophy of non-duality, which holds that what appears to be these two different things are ultimately both the same thing. The relationship of ultimate singularity and identity between these two poles of Existence is not even possible to grasp conceptually. This is because everything we know or conceive is experiential in nature, and so must appear as either this or that, e.g., as wave or particle, as opposed to being this and that simultaneously. However, What's Actually There as these two poles of Individuality is not an experience. Rather, experience is our apprehension of something created as a result of a relation occurring between What's Actually Here and What's Actually There, and in all cases What's Actually Here and There is the same indivisible Existence functioning as different poles of Individuality. Thus, although what's Actually Here and There is indivisible, when conceiving of Itself, What's Actually Here and There must appear to Itself, from the Individual perspective, as either this or that, i.e., as either the Individual or the More Fundamental Individuality, i.e., as one pole or the other of this experientially created duality, when in actuality it is both simultaneously.

There is only Existence, indivisible and non-dual, i.e., not actually two. All duality, all apparent separateness and divisibility, is an artifact of experience, an artifact of how experience is created as the product of some relation of Existence to Itself, as that product is apprehended by Existence on one side of that relation. Thus, when indivisible Existence conceives of Itself it must do so through the dualizing and polarizing lens of experience, and through that experiential lens Existence must appear to Itself as dual, as two different and therefore seemingly separate things, i.e., as the Individual and as the More Fundamental Individuality, as the Atman and Brahman.

Because what we apprehend as experience must be created through some relation of Existence to Itself in order to exist, in which relation we ourselves are always involved as one of the poles of Existence involved in that relation, and because what we apprehend as experience is that creation as it appears from whatever side of the relation we are on, experience is limited to being or presenting itself as either this or that. Thus, experience says if something is this, then it is not that, if it is white then it is not black, if it is up then it is not down, and so on. And this mutual exclusivity of being between opposites is true of experience. However, this mutual exclusivity of being between opposites is not true of Existence, because the limitations of experiential nature are not limitations that inhere in the Nature of that which is Itself the source of experience. That is, unlike experience, Existence is not created and so is not limited in the way experience is limited.

That having been said, as much as Existence is indivisible, i.e., non-separable from Itself, it is also Individual. That is, the Reality of Existential Individuality is just as valid and real as the Reality of Existential Indivisibility or Oneness. Put another way, Existence is simultaneously both One and Individual, i.e., Indivisible and Individual. Experience tells us that it has to be one or the other, but experience is wrong in this regard, because for reasons just explained, experience cannot itself convey the actual Nature of Existence, as that Nature is non-experiential, as it is that Nature that is Itself the source and basis of all experience and so of all perceived and conceived duality.

Spirituality tends to stress or focus upon the Indivisible aspect of Existence, upon the ultimate Oneness of everything, as a counterbalance to our usual materialistic focus upon our Individuality and the apparent differences between Individuals that seem to separate this Individual from that Individual. However, in focusing upon the Reality of Existential Indivisibility, spirituality tends to assign a lesser status to the equally valid Reality of Existential Individuality, in as much as spirituality often treats Existential Individuality as if it were ultimately unreal or an illusion. However, this treatment of Existential Individuality as unreal in the context of a focus upon Existential Indivisibility is itself an artifact of experience, because when the focus is upon one aspect of Reality the focus cannot be on the other, experientially opposite, aspect of Reality, in which case the other aspect then seems either unreal or less real, or somehow subordinate to the one aspect. And so when God is considered as the Reality, the Individual must then seem less real or somehow subordinate to that Reality, when that is not the actual relation between these two aspects of Existence.

In the same way, science, in its focus upon physical experiential reality, must view the existence of the very thing that apprehends experience, i.e., Consciousness, as a lesser or subordinate reality, as a created reality, as somehow being produced through the machinations of material reality, even though it is material reality that is itself produced through the machinations of Consciousness-Existence. Again, when the one is experienced as real, the opposite must be experienced as less real or unreal, or as somehow subordinate, regardless of their actual relation.

Thus, even though it is not possible to conceive of Existence as being simultaneously Individual and Indivisible, owing to the unavoidable experiential limitation that makes it impossible for an Individual to be simultaneously involved in the mutually exclusive relations necessary to create and apprehend opposite or complementary experiences, that is nonetheless its Nature.

There is only Existence and experience. There is only that which Exists and what that which Exists creates as reality as a result of its relations to Itself. However, that which Exists and which through relation to Itself creates experience, does not create experience en masse, but rather does so at the level of the Individual. That is, what the Individual apprehends as experience is not apprehended by the whole of Existence as experience, but is apprehended as experience only by the Individual point of Existence, the Individual point of Consciousness, that is apprehending the experience. Thus, what I experience is what I experience, according to the relations with Existence in which I, as an Individual, am involved, and what you experience is what you experience according to the relations with Existence in which you, as an Individual, are involved. And it is the same for every Individual, including the More Fundamental Individuality, which is also creating experience as an Individual, according to the relations with Existence in which it is involved.

Thus, every Individual, regardless of scale, creates their own reality, their own unique set of experiences, according to how they are being in relation to the rest of Existence, and specifically, according to how they are being in relation to their More Fundamental Individuality. And since, owing to the indivisible nature of Existence, every Individual is inseparable from and so part of every other Individual's More Fundamental Individuality, how one is being in relation to other Individual's is also how one is being in relation to their More Fundamental Individuality.

Most importantly, how any Individual is being in relation to its More Fundamental Individuality is something that is determined by each Individual, according to what that Individual is, in that moment, choosing as its *mode of being*. The Individual's mode of being is nothing more than the Individual's in the moment attitude toward, and relation to, its More Fundamental Individuality, which attitude and relation is always one of either *allowing* or *resistance,* and which attitude is always chosen by the Individual.

Further, no Individual, regardless of scale, can force any other Individual to choose to be in one mode of being rather than the other, as that choice always is always one that every Individual is, in every moment, free to make themselves, which is why the Individuals inherent ability to make that choice is called *free will*, i.e., the Individuals ability to freely choose how it will be in relation to what is ultimately Itself. And because no other Individual, regardless of scale, can force any other Individual to choose to be in one mode of being rather than the other, and because what the Individual creates and apprehends as experience in any moment is determined by what the Individual is, in that moment, choosing as their mode of being, no other Individual, regardless of scale, can force or impose or assert any experience upon any other Individual, because everything an Individual creates and apprehends as experience is a product of how the Individual that is apprehending the experience is Itself, according to its own exercise of free will, choosing to be in relation to its More Fundamental Individuality.

If there were only Indivisibility and not Individuality there would be no experience, because in the absence of the Reality of Individuality there is no basis for Existential Self-relation. All Existential Self- relation is a relation of an Individual to a More Fundamental Individuality, and every More Fundamental Individuality is Itself an Individual relative to another More Fundamental Individuality, ad infinitum, because Existence is infinite and so without boundaries within or without. The concept of wholeness does not apply to Existence, because wholeness implies some external boundary, and such a concept is counter to that of the Reality of

Existential infinity. However, the concept of Indivisibility can be applied to Existence because that concept does not impose limits upon Existence, as long as one remembers that Existential Indivisibility does not preclude the simultaneous Reality of Existential Individuality, including the free will inherent in each and every point of Existence, regardless of scale.

Every Individual, regardless of scale, creates their own reality, their own set of experiences. Period. There are no exceptions because all Individuals are ultimately the same indivisible Existence, and so all have the same inherent ability to choose, in each moment, their own mode of being regardless of what any other Individual is choosing as their mode of being. In other words, each and every Individual, regardless of scale, is *autonomous* with regard to what it is choosing as its mode of being, and so is autonomous with regard to how it is being involved in the relations with Itself that creates what it then apprehends as experience.

To some degree Existence is like a River that is free to choose its direction of flow, and where every Drop in the River is also free to choose its own direction of flow regardless of what the River is choosing as its direction of flow. And even if every Drop chooses to flow in a direction that is the opposite of the direction the River is choosing to flow, the River still flows in the direction it has chosen, and the Drops still flow in the direction they have chosen. This cannot be grasped, but that is how Existence operates in the creation of experience at the level of the Individual, which is according to how the Individual is themself freely choosing to flow in relation to their More Fundamental Individuality. The River cannot and does not tell the Drops in which direction to flow, and the Drops cannot and do not dictate the direction of flow of either the River or the other Drops. But each Drop must, in each moment, choose and so determine its own direction of flow, and in so doing choose and determine its relation to the River, which relation, as will be described shortly, is the basis of any and all experiential wantedness or unwantedness, happiness or unhappiness, that any Drop, i.e., Individual, ever creates and apprehends.

Regardless of whether or not an Individual is or is not aware of their role in the creation of what they apprehend as experience, and regardless of whether or not an Individual is or is not aware that they are one pole of an indivisible duality of Existence, all experiential wantedness and unwantedness apprehended by an Individual is nonetheless created as a function of how that Individual is choosing, in each moment, to be in relation to the other pole of their Existence, i.e., to their More Fundamental Individuality. And as previously stated, what the Individual is choosing through their exercise of free will is their mode of being, and that mode of being is either allowing or resistant. And that chosen mode of being as either allowing or resistant places the Individual in a relation of either aligned or oppositional flow, respectively, relative to their More Fundamental Individuality, and it is that relation of either aligned or oppositional Existential flow that is the basis of all experiential wantedness or unwantedness, respectively, that is created and apprehended by any Individual, regardless of scale.

Specifically, when an Individual chooses to be in a mode of being that is allowing, that Individual is, in that moment, involved in a relation of aligned flow with their More Fundamental Individuality, in which case the experiences that Individual creates and apprehends in that moment as a result of their involvement in that relation have the quality of wantedness.

Conversely, when an Individual chooses to be in a mode of being that is resistant, that Individual is, in that moment, involved in a relation of oppositional flow with their More Fundamental Individuality, in which case the experiences that Individual creates and apprehends in that moment as a result of their involvement in that relation have the quality of unwantedness. Ultimately, experiential wantedness and unwantedness are the experiential result and product of our choice, as Individuals, to either flow with or against, respectively, the other pole of our own Individuality. Thus, the attractive quality of wantedness and the repulsive quality of unwantedness are like the difference between what is felt when swimming downstream or upstream, respectively, in which case one's own direction of flow is either augmented by or resisted by the river's own direction of flow. However, here it is important to note that if you feel unwantedness it is not because the River is choosing to flow in resistance to you, rather, it is because you are choosing to flow in resistance to the River.

Thus, as previously stated, all experiential wantedness and unwantedness is ultimately a function of how the Individual is choosing to be in relation to their More Fundamental Individuality. But when an Individual is unaware of the presence or functioning of this relation, unaware of their True Nature as that which both creates and apprehends experience, it must then seem to that Individual as if the wantedness and unwantedness they experience is inherent in the experiences themselves, inherent in the objects of experience. Put another way, when one's True Nature is hidden, then the Creator of experience is also hidden, making experience appear to be what's actually there, when What's Actually There is the non-experiential Consciousness-Existence that, through relation to Itself, is creating and apprehending the experience. This is analogous to what happens if one sees what is only a reflection, but is unaware of the presence of a mirror, in which case the reflection appears to be what's actually there, when what's actually there is the mirror.

And once it seems that experience is what's actually there, it also seems that wantedness and unwantedness are inherent in experience itself, in which case the Individual still spends their time trying to do what is in their Nature, which is trying to create a wanted rather than unwanted experience. However, once it seems that wantedness and unwantedness are inherent in experience itself, rather than something created by the Individual that is apprehending the experience, the Individual then acts in accord with their Nature by pursuing those things that they think will, once they possess them, cause them to feel happy, and pushing away those things that they think will, if they do possess them, cause them to feel unhappy. And in this way the Individual chooses unconsciously and unknowingly its mode of being, and so chooses unconsciously and unknowingly the wanted or unwanted quality of the type of experiences it creates and apprehends.

What happens in this method of trying to create experiential wantedness, in unawareness of the actual relation that is responsible for the creation of those experiential qualities, is that one often chooses to be in a mode of resistance in order to either pursue or push away those objects that they think will make them happy, in which case the Individual inadvertently creates for themself unwanted rather than wanted experience while in pursuit of that which is wanted.

The reason wanted objects seem to make us feel good is because we find them easy to allow, i.e., we reflexively allow them, and so easily enter into a relation of aligned Existential flow, which relation then creates experiential wantedness. And the reason unwanted objects seem to make us feel bad is because we find them easy to resist, i.e., we reflexively resist them, and so easily enter into a relation of oppositional Existential flow, which relation then creates experiential wantedness. Again though, in the absence of realizing that this relation between the Individual and the Inner Self is always, in every moment, operant, and is what is actually producing the wantedness and unwantedness associated with any experience, it must then seem that those qualities of wantedness and unwantedness inhere in the objects of experience themselves, which appearance leads to what is referred to in Eastern philosophies as *attachment* and *aversion*, which is simply the reflexive clinging to experiences that seem to have a quality of wantedness and the reflexive pushing away of experiences that seem to have a quality of unwantedness, respectively, both of which actions have the Individual choosing, at an unconscious level, to be in a mode of resistance, thereby placing the Individual in a relation of oppositional Existential flow with respect to their More Fundamental Individuality, thereby creating for that Individual an unwanted rather than wanted experience.

In the pursuit of happiness one must see happiness as something to be obtained, as something that exists independent of the Experiencer of it, and this is not the actual nature of happiness. The actual nature of happiness is that it must be and always is created by the Individual that is apprehending it, according to a specific relation in which the Individual is involved with their More Fundamental Individuality. And if an Individual is not choosing to be involved in the relation that creates happiness, then they are choosing to be involved in the relation that creates the opposite of happiness, for although it is true that each Individual gets to choose the nature of their relation to their More Fundamental Individuality, and so gets to choose whether they create in each moment wanted or unwanted experience, it is also true that each Individual has no choice but to choose in each moment to be in one relation or the other with their More Fundamental Individuality, i.e., in a relation of aligned or oppositional Existential flow, and so must in each moment create either wanted or unwanted experience.

And so pursue happiness if you must, but know that there is an easier way, and a more effective way, which is to not pursue it, but to just create it, by choosing in any moment to allow rather than resist, to let be rather than push against. Just try it and see how it feels, i.e., see whether you create and apprehend in the moment of allowing or resistance emotional wantedness or unwantedness, as emotions are the most fundamental experience and so are the most directly and immediately reflective of the aligned or oppositional relation of the Individual to their More Fundamental Individuality, which relation is itself a direct result of what the Individual is choosing as their mode of being in that moment.

Everything we want we want because we think that in having it we will feel better, that as a result of getting what we want that we will experience a more wanted emotion. This is true for both the saint and sinner, as both are moved inexorably by their Nature to create wanted rather than unwanted experience. Thus, the sage pursues enlightenment for the same reason the junkie puts a needle in his arm. There are an infinity of Individuals, but all are composed of the same Existence, and thus all have the same Nature. And yet because they are Individuals, even though they are all acting according to the same Nature, they each produce different experiential results because they are each acting according to that same Nature according to their own exercise of free will, according to how they and they alone are choosing to be in relation to the other pole of their own Individuality as they seek to create for themself an always more wanted emotional experience. Put another way, we all want the same thing because we are all the same non-experiential Thing, but because we are also Individuals we each have our own way of moving toward that thing that we all want.

Unaware of our True Nature and so of our role in the creation of what we apprehend as experience, we try to create wanted emotion by arranging external reality in a way that we can reflexively allow. However, in getting that arrangement in place we often choose to be in a mode of resistance, and so create an experience that is the opposite of the experience we wanted to create, i.e., we create unwanted rather than wanted emotion. Also in our unawareness of our True Nature we often forget why it is that we wanted the object in the first place, which is always for the same reason, i.e., to have a more wanted emotional experience, in which case we continue our pursuit of the object regardless of how our pursuit of it is making us feel.

However, if you can realize that the goal is always a more wanted emotion, the creation of a wanted rather than unwanted emotional experience, and that that goal can be reached regardless of the arrangement of external reality, as what is apprehended as emotion is actually a function of a different relation, then it is easier not to cling to the wanted object of experience, or push against the unwanted object of experience, which then makes it easier to choose to remain or become involved in the relation of Existential alignment that actually creates wanted experience. But if you forget or don't realize what it is that you are actually always after, and forget or don't realize the actual cause of experiential wantedness and unwantedness, then clinging to the wanted object and pushing against the unwanted object seem to be the only viable options to get what you want, in which case one then becomes trapped in a self-perpetuating loop, inadvertently creating unwanted experience while in the pursuit of wanted experience.

This is not a world of suffering. To the contrary, it is a world where it is just as easy to create the opposite of suffering as it is to create suffering. One just needs to know which button to push on the machine of experiential creation. However, because most people are unaware of both how experience is created, as well as what it is they actually want, which is just to feel better, to create and apprehend a more wanted emotional experience, most people go about trying to create experience in a way that is the opposite of the way it is actually created, leaving them pushing the button of resistance rather the button of allowing, in which case they end up producing experiential results that are the opposite of those they intended to produce.

But as they say, it's all good, because even in the inadvertent creation of unwantedness, the purpose for which we came is still being served, as we serve that purpose regardless of whether or not we are cognizant of our True Nature and regardless of the degree to which we create wanted or unwanted experience while we are here. It's just that it's usually more enjoyable to serve that purpose when one understands the rules of the game we are playing, regardless of whether or not one is cognizant of the purpose of the game itself, which purpose is a story for another time.

To learn more about how physical experience is created, as well as the limitations inherent in the creation of experience, I recommend my article, *The Experiential Basis of Wave-Particle Duality and The Uncertainty Principle*, published in the Prespacetime Journal, Vol 2, No 4 (2011).

To learn more about how Existence evolves into that which underlies what we apprehend as the world around us as a result of becoming involved in the progressive relations with Itself that create what it and we apprehend as emotional, mental, and physical experience, I recommend my series of four articles collectively titled; *Existential Mechanics: How the Relations of Existence to Itself Create the Structure of Reality and What We Experience as Reality*, recently published in the Journal of Consciousness Exploration and Research, Vol 2, No 9 (2011).

Article

Existential Cause & Experiential Effect

Steven E. Kaufman*

ABSTRACT

The idea that what we experience as physical-material reality is what's actually there is the flat Earth idea of our time. That is, the idea that physical-material reality is what's actually there where we experience it to be is an idea that, based upon appearances, seems to be true, in the same way that while standing in the middle of Illinois the Earth appears to be flat, but from a broader perspective is seen to be but an illusion of limited perspective. That broader perspective is afforded by the limitations of experience revealed by quantum physics in the form of the phenomena of wave-particle duality, quantum uncertainty, and quantum entanglement, which limitations, in revealing the nature of experience to be Experiencer dependent, provide insight into the way in which experience is created as the product of a relation of Consciousness, i.e., What Is Actually There, to Itself. However, the same limitations of experience revealed by these phenomena serve to hide from view what these phenomena reveal about the nature of experiential reality, including how experiential reality is created, when considered within a materialistic framework, i.e., within a framework wherein material reality is conceived of as being what's actually there. Thus, although it may seem that we live in a world of material cause and effect, we actually live in a world of Existential cause and experiential effect. That is, we live in a world where the cause is always some relation of Consciousness-Existence to Itself, and the effect is always the experience that is created and apprehended by the Individual Consciousness involved in that relation.

Key Words: existential cause, experiential effect, Existence, Consciousness, material world.

We do not live in a material world. That we live in a material world is an illusion. The material world is an experiential world, and as such it is a reflection that arises within and rests upon the Mirror of What Actually Exists, and it is in the world of that Mirror that we actually live, whether we know it or not. However, the material world is not itself an illusion, as it exists as a reality, i.e., as an experiential reality, as a reflection exists on the surface of a mirror. The illusion is the thought that material reality is what actually exists where it appears to be, the illusion is the thought that material reality is what's actually there where it appears to be, in the same way that it is an illusion to think that a reflection is what's actually there where it appears to be, since what's actually there is whatever it is upon which the reflection rests and within which it arises.

In the case of the reflection-experience that is material reality, what's actually there upon which that reflection rests and within which it arises is Consciousness-Existence, i.e., that which through relation to Itself both creates and apprehends experiential reality. And so the materialists have it backwards, which is to say, they see the relation between material reality and Consciousness in a way that is the complete opposite of their actual relation. That is, materialists

*Correspondence: Steven E. Kaufman, Independent Researcher. http://www.unifiedreality.com
E-mail: skaufman@unifiedreality.com

see material reality, or some version of material reality, e.g., quantum reality, as producing Consciousness through some sort of material cause and effect, wherein material reality is the cause and Consciousness the effect.

Because materialists take material reality in one form or another for what's actually there, they are unable to recognize Consciousness as what's actually there, just as when one takes a reflection on the surface of a pond for what's actually there the pond becomes hidden. It is in this context, in this experiential framework, that it must seem to the materialist that material reality is the cause and Consciousness the effect, when again, their relation is the exact opposite, i.e., Consciousness is the cause and material reality or experiential reality, is the effect.

Consider that you were raised in a world where you were taught that reflections were the reality, were what's actually there, and then at some point you become cognizant of a mirror. What then are you to make of the mirror and of its place in reality? The position of actuality, of cause, is already occupied, and so the mirror must somehow be crammed into the position of effect. This is what occurs in the materialist view of reality, wherein one attempts to account for Consciousness within a framework where material reality is taken as causal, taken for what's actually there. That is, Consciousness is seen as effect not because it is effect, but because that is how it must be seen within a materialistic framework, within a framework where material reality is seen as causal. It is as if one spent their life thinking that a board was the causal reality, and then they come across a tree and, still holding to the idea of the board as causal, they then go about trying to figure out how the tree comes from the board.

We understand the absurdity and futility of trying to figure out how a tree comes from a board, because we understand their cause and effect relation. Materialists however do not understand the absurdity and futility of trying to figure out how Consciousness comes from material-experiential reality, because what they understand as their cause and effect relation is the exact opposite of their actual cause and effect relation. When an Individual sees what's up as down, that Individual must then see what's down as up. And when an Individual conceives of effect as cause, that Individual must then conceive of cause as effect. This linkage in the way an Individual must apprehend what are opposite or complementary experiences is a function of an experiential limitation I call *experiential entanglement*, which limitation, like all experiential limitations, is a function of the fact that all experience is the product of a relation in which the Individual Consciousness that is apprehending the experience must themself be involved.

That all experience is the product of a relation in which the Individual that is apprehending the experience must themself be involved, along with the fact that opposite or complementary experiences are always the product of opposite and so mutually exclusive relations, imposes some limitations upon what it's possible for an Individual to create and apprehend as experience in any one moment. One of those limitations is that it's not possible for an Individual to be simultaneously involved in the mutually exclusive relations necessary to create opposite experiences. I call this limitation the *principle of the preclusion of an Individual's simultaneous creation and apprehension of experiential opposites* or, more succinctly, the *experiential preclusion*. It is this experiential limitation, this experiential preclusion, that is responsible for the phenomena of wave-particle duality and quantum uncertainty, since this experiential limitation dictates that for any experience that an Individual creates there is an opposite experience that Individual cannot create in that same moment, because creating that opposite experience would require the Individual's involvement in a relation that is mutually exclusive of the relation in which the Individual must presently be involved in order to create what they are already, in that moment, apprehending as experience. Thus, if an Individual Consciousness is involved in a relation with an Underlying Actuality, which is also Consciousness, that creates what that Individual apprehends as a particle experience, that Individual cannot, in that same moment, be involved in the mutually exclusive relation with that Underlying Actuality necessary to create a wave experience. Opposite or complementary experiences are always the product of opposite and so mutually exclusive relations, and it's not possible for an Individual to be simultaneously involved in mutually exclusive relations, just as its not possible for an Individual to simultaneously face North and South, since facing one direction means you are not facing the other.

However, this experiential limitation, this experiential preclusion, does not just operate in the creation of quantum experience, rather, it operates in the creation of experience at all levels, emotional, mental, and physical. At the emotional level it is the experiential preclusion that makes it impossible for you to feel good when you feel bad, and vice versa, as positive and negative emotions, wanted and unwanted emotions, being opposite experiences, are the products of opposite and so mutually exclusive relations. At the mental level it is the experiential preclusion that makes it impossible to know the Earth as round as long as you know it to be flat, to believe in evolution while believing in the biblical version of events, or to know Consciousness as what's actually there while knowing material reality to be what's actually there. We are not generally aware of the functioning of this experiential limitation, this experiential preclusion, because what it does is create an experiential blind spot with regard to whatever experiences are the opposite of those you are presently creating and apprehending as reality. And what is a blind spot but a place you don't know that you can't see because it already seems to you that you are seeing what's there.

There is another limitation upon what it's possible for an Individual to create as experience owing to the fact that all experience is the product of a relation in which the Individual that is apprehending the experience must themself be involved, which limitation is the corollary of the experiential preclusion just described. The experiential limitation that is the experiential preclusion has to do with what it's not possible for an Individual to apprehend as experience owing to the impossibility of that Individual being simultaneously in mutually exclusive relations, e.g., facing North and South simultaneously. The other experiential limitation, which I refer to as experiential entanglement, has to do with the way in which an Individual must create experience through relations that are *mutually inclusive* of the relations in which they are already involved, mutually inclusive of the relations in which they must be involved in order to create what they are presently creating and apprehending as experience.

Thus, one experiential limitation involves what an Individual can't create as experience according to mutually exclusive relations in which they can't be simultaneously involved, while the other experiential limitation involves what an Individual must create as experience according to mutually inclusive relations in which they must be simultaneously involved. And both of these limitations have as their basis the fact that all experience, rather than being something that just sits there waiting for us to happen across, is the product of a relation in which the Individual that is apprehending the experience must themself be involved, which necessary involvement of the Individual in some relation in order to create what they apprehend as experience then imposes upon that Individual limitations regarding other relations in which they can become involved as long as they continue to remain involved in a particular relation in which they create and apprehend a particular experience.

Every particular experience that an Individual apprehends is the product of a particular relation in which that Individual must be involved in order for them to create and apprehend that particular experience. Therefore, as long as an Individual continues to have a particular experience they must remain involved in the particular relation that creates for them that particular experience, and the necessity of their being in that particular relation in order to continue to create that particular experience imposes upon that Individual two related limitations with regard to other relations in which they can become involved in order to create other experiences, one of which is a limitation imposed by the impossibility of the Individual being involved simultaneously in mutually exclusive relations, and the other of which is a limitation imposed by the necessity of the Individual's simultaneous involvement in mutually inclusive relations. The experiential limitation involving mutually exclusive relations, i.e., the experiential preclusion, dictates what it's not possible for an Individual to create and apprehend as experience according to what that Individual is presently creating and apprehending as experience, whereas the experiential limitation involving mutually inclusive relations, i.e., experiential entanglement, dictates the way in which an Individual must create and apprehend experience according to what that Individual is presently creating and apprehending as experience.

Both of these limitations, i.e., the experiential preclusion and experiential entanglement, are functioning at all times in the Individual's creation of experience at every level of experience, emotional, mental, and physical, as well as between levels of experience. As already stated, it is the experiential preclusion that makes it impossible to feel good while feeling bad, and vice versa. However, it is experiential entanglement that seems to color all other experience with wantedness or unwantedness when one is feeling good or bad, respectively. How many poems and songs have been written about how when one falls in love all the world is suddenly brighter, or how when love is lost all the world is suddenly dark? Such associations between different experiences are the result of experiential entanglement, i.e., the necessity of the Individual's involvement in what are mutually inclusive relations as they create what they apprehend as experience in any one moment. To feel love, a very positive and wanted emotion, one must be in a relation of Existential alignment, whereas to feel the opposite, a very negative and unwanted emotion, one must be in a relation of Existential opposition. The experiential preclusion dictates that if you are in one relation then you are not in the other, as these relations are mutually exclusive. Experiential entanglement dictates that whichever relation you are in, i.e., Existential alignment or opposition, then all other relations in which you become involved in that same moment as you create mental and physical experience must be mutually inclusive of that relation, meaning they must be relations that have the same aligned or oppositional orientation, and so must be created as experiences that have the same quality of wantedness or unwantedness as that of the emotional experience that is also being created in that moment.

Also as already stated, it is the experiential preclusion that makes it impossible to conceive of the Earth as being round while conceiving of it as being flat, as those are opposite experiences that must then be the product of what are mutually exclusive relations. However, it is experiential entanglement that dictates that as long as one conceives of the Earth as being flat then the idea of a round Earth must be seen as false or unreal, because as long as one is creating and apprehending the mental experience-concept of the Earth as being flat then the only way to simultaneously conceive of the Earth as round is through a relation that is mutually inclusive of the relation in which the Individual is already involved as they create for themself the idea-experience of the Earth as being flat, which mutually inclusive relation is one that creates the idea-experience of the Earth as not-being round.

The Earth cannot be conceived of as being both flat and round simultaneously by a single Individual, as those are opposite concepts and therefore limited in their creation by the experiential preclusion. But the Earth can be conceived of as being flat and not round simultaneously, because those are not opposite concepts, as they are derived from what are mutually inclusive relations. And owing to experiential entanglement, if the Earth is conceived of as being flat, if that is the idea that is being held to, if that is the idea that the Individual is actively creating, then from that perspective, from within that relational framework, the idea of the Earth's roundness must be conceived of as being false. Thus, one experiential limitation dictates what cannot be created simultaneously as experience by an Individual according to what that Individual is already creating as experience, while the other experiential limitation dictates what an Individual must create as experience according to what that Individual is already creating as experience. Put another way, in terms of relations, one experiential limitation, i.e., the experiential preclusion, dictates the mutually exclusive relations in which an Individual cannot become involved in order to create experience according to the relations in which that Individual must already be involved in order to create what they are presently apprehending as experience, while the other experiential limitation, i.e., experiential entanglement, dictates the mutually inclusive relations in which an Individual must become involved in order to create experience according to the relations in which that Individual must already be involved in order to create what they are presently apprehending as experience.

And this then brings us back to Existential cause and experiential effect, and to the unavoidable reversal of the actual relation between Consciousness and experience, wherein experience must be conceived of as cause and Consciousness as effect, by any Individual that holds to the idea of material reality as being what's actually there, in which context material reality must, according to experiential entanglement, be seen as causal, and in which context, also according to experiential entanglement, the actual cause, i.e., Consciousness, must then be seen as effect. Put another way, materialists can't help but conceive of Consciousness as an effect of material reality owing to the limiting effect of experiential entanglement, which limiting effect dictates that Consciousness, if it is to be apprehended at all, must be apprehended from a relation that is mutually inclusive of the relation that creates the idea of material reality as casual, from which relational framework Consciousness must then be viewed or seen as effect. When up is seen as down, if down is to be seen at all, it must be seen as up, and when effect is conceived as cause, if cause is to be conceived at all, it must be conceived as effect. That is experiential entanglement, which, like all experiential limitations, is a function of the fact that what we experience as reality is not there as we experience it to exist independent of our experience of it as such, but rather only exists as we experience it to exist according to some relation in which we, as Individuals, are involved with What Is Actually There, understanding that What Is Actually There is not different or other than What Is Actually Here where we are, both of which are non-experiential Consciousness-Existence.

And so, owing to experiential entanglement, as long as we see material reality as being what's actually there it must also seem that we live in a world of material cause and effect, although we really live in a world of Existential cause and experiential effect, a world where Consciousness, through its relations to Itself, is always the cause and experience is always the effect.

The problem for idealists, i.e., those who consider Consciousness to be primary or casual, has been explaining how the somethingness of material and experiential reality can be produced by the non-experiential Reality of Consciousness. The missing link has been with regard to how it is that Consciousness-Existence creates experience, and so creates what we, as Individual points of Consciousness, apprehend as material reality in particular and experiential reality in general. However, that missing link has been found and it is as follows: Consciousness-Existence creates experience by being in relation to Itself, because as a result of any relation of Consciousness-Existence to Itself something is created that is not Consciousness, which created something the Individual Consciousness involved in that relation apprehends, from its perspective within that relation, as experience, as an experiential reality. The actual relations between all these different concepts are shown in the drawings below.

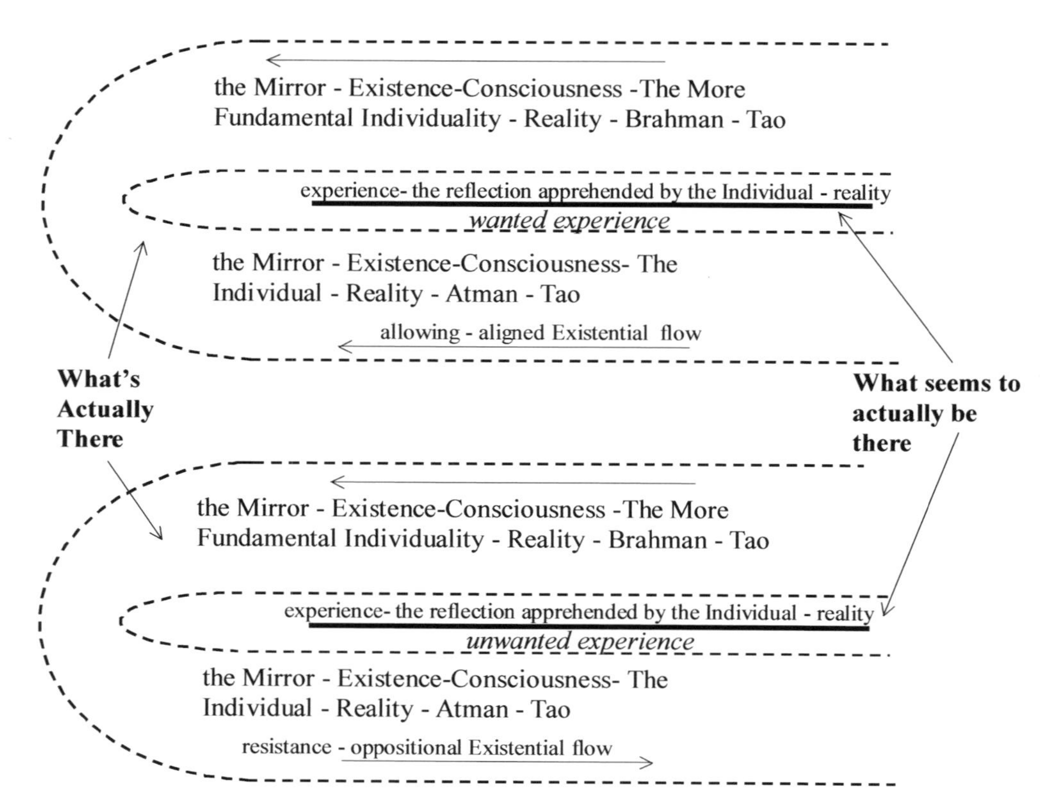

Figure 1 These two drawings each depict a sort of cross section of Consciousness-Existence being in relation to Itself and as a result creating what it then, from the perspective of the Individual, apprehends as experience. The dashed lines represent What Actually Exists, i.e., Existence-Consciousness-Reality, etc., while the solid line represents that which What Actually Exists creates as a result of its relation to Itself, which creation is then apprehended from the perspective of the Individual as an experiential reality, which experiential reality, like a reflection that rests within a mirror, can be taken, i.e., mistaken, for what's actually there, in which case, owing to experiential entanglement, What's Actually There as Cause must then appear to only seem to exist as effect, if it is seen to exist at all. The drawing at the top depicts a relation of aligned Existential flow, i.e., a relation in which the Individual is choosing, via its exercise of free will, to project Itself in alignment with the flow of its More Fundamental Individuality, thereby creating for Itself an experience-reflection that is apprehended as having a wanted quality, while the drawing at the bottom depicts the opposite, mutually exclusive relation of oppositional Existential flow, i.e., a relation in which the Individual is choosing, via its exercise of free will, to project Itself in opposition to the flow of its More Fundamental Individuality, thereby creating for Itself an experience-reflection that is apprehended as having an unwanted quality.

And because anything that an Individual apprehends as experience must be created as a result of some relation with Existence in which the Individual that apprehends the experience is themself involved, and because an Individual cannot choose to flow simultaneously both in alignment with and opposition to Itself, as those are mutually exclusive relations, an Individual cannot simultaneously create and apprehend both wanted and unwanted experiences. That is one limitation upon an Individual's creation of experience, limiting what an Individual can create and apprehend as experience in any moment according to the relations in which that Individual must already be involved in order to create what that Individual is already apprehending as experience. And since an Individual cannot simultaneously be involved in the mutually exclusive relations necessary to create opposite experiences, this then means that in any one moment whatever relations in which an Individual is involved in order to create what that Individual is apprehending as experience must be mutually inclusive relations. This is the other limitation upon an Individual's creation of experience, dictating what an Individual must create and apprehend as experience in any moment according to the relations in which that Individual must already be involved in order to create what that Individual is already apprehending as experience. Thus both limitations serve to restrict what an Individual can, in any one moment, create and apprehend as experience based upon other relations in which that Individual is already involved as it creates what it is already, in that moment, apprehending as experience. However, one limitation is negatively restrictive, whereas the other is positively restrictive, as the former dictates what cannot be created as simultaneous experience by a single Individual, whereas the latter dictates what must be created as simultaneous experience by a single Individual. Wave-particle duality and quantum uncertainty are negatively restrictive experiential phenomena that have as their basis the negatively restrictive experiential limitation referred to as the experiential preclusion, whereas quantum entanglement is a positively restrictive experiential phenomenon that has as its basis the positively restrictive experiential limitation referred to as experiential entanglement.

The experiential limitations that manifest so vividly and paradoxically at the quantum level are happening at every level of experience, with regard to every experience we create, it's just that we don't recognize the moment to moment operation and functioning of these limitations owing to our complete immersion in the experiential reality, in the reflection, we are, through our relations to the rest of Existence, creating. Quantum phenomena are only paradoxical in the context of a materialistic framework, in the context of a conception of reality where material reality is apprehended as causal. Conversely, in the context of an idealistic framework where material and quantum reality are seen as effect, there is no paradox, rather, there is instead the expected result of limitation owing to the relations necessary for the Cause to create the effect. Of course if you think that things are as they are regardless of your experience of them as such it then must seem strange and paradoxical that something could appear as either wave or particle. But if you realize that things only are as they are according to your involvement in the relation that causes you to apprehend them as such, as a particular experience, then it is not paradoxical that while in one relation one appearance-experience would be created and while in the opposite relation the opposite appearance-experience would be created. It also seems paradoxical in the context of a materialistic and therefore mechanistic framework that having one experience could somehow instantaneously, and so outside the boundaries of any possible material mechanism, influence what else is experienced. But again, if you realize that things only are as they are according to your involvement in the relation that causes you to apprehend them as such, it is not paradoxical that being involved in the relation that creates one experience dictates what other relations are possible for you in that same moment and so dictates what else can be created as experience by you in that moment. The difference between paradox and understanding lies in whether one sees experience as being what's actually there, be it either a gross material or more subtle quantum experience, or whether one sees experience as a reflection that arises upon and rests within something that is completely and utterly non-experiential, and yet is Itself the basis of all experience.

Thus, this explanation of the nature of Reality and reality, the nature of What's Actually There and what seems to actually be there, is not an explanation devoid of science. To the contrary, it is an explanation that rests upon the furthest reaches of science, as it rests upon the limitations of experience encountered as scientists have tried to quantify and examine the smallest bits of material reality, i.e., it rests upon the phenomena of wave-particle duality, quantum uncertainty, and now upon the phenomenon of quantum entanglement as well. Scientists have not yet figured out the basis of these phenomena because they continue to look at them within a materialistic framework, i.e., within a framework where material reality is still seen as primary and therefore causal. And science will never, be it another hundred or a thousand years, find an explanation for these phenomena within a materialistic framework, because these phenomena have no explanation from within that framework, because these phenomena are the not the product of any material cause and effect relation, rather they are the product of an Existential cause and experiential effect relation, and it is only within that framework that their basis can actually be explained.

Is it possible to explain how a tree comes from a block of wood? It is certainly possible to try. But is it possible that such an explanation will ever have any actual validity, since the very basis of the explanation is based upon an inversion of the actual cause and effect relation between the objects in question? No. Is it possible to come up with a material or quantum reality based mechanical explanation for wave-particle duality, quantum uncertainty, and quantum entanglement, as well as Consciousness? It is certainly possible to try, as science has demonstrated. But is it possible that such an explanation will ever have any actual validity, since the very basis of the explanation is based upon an inversion of the actual cause and effect relation between the objects in question? No. There are many scientists who have understood that these phenomena indicate that Consciousness must be part of the equation, but there are few if any who understand that in that equation it is Consciousness Itself that is completely causal and material reality, experiential reality, that is purely the effect, because as scientists they operate within a conceptual framework of objectivity and material causality, which, owing to experiential entanglement, makes it impossible for them relegate to the position of pure effect that which they experience as reality.

The idea that what we experience as physical-material reality is what's actually there is the flat Earth idea of our time. That is, the idea that physical-material reality is what's actually there where we experience it to be is an idea that, based upon appearances, seems to be true, in the same way that while standing in the middle of Illinois the Earth appears to be flat, but from a broader perspective is seen to be but an illusion of limited perspective. That broader perspective is afforded by the limitations of experience revealed by quantum physics in the form of the phenomena of wave-particle duality, quantum uncertainty, and quantum entanglement, which limitations, in revealing the nature of experience to be Experiencer dependent, provide insight into the way in which experience is created as the product of a relation of Consciousness, i.e., What Is Actually There, to Itself. However, the same limitations of experience revealed by these phenomena serve to hide from view what these phenomena reveal about the nature of experiential reality, including how experiential reality is created, when considered within a materialistic framework, i.e., within a framework wherein material reality is conceived of as being what's actually there.

At this point I would like to make very clear that none of this, in anything that I have written or will write regarding this subject, is meant as a criticism of Individual scientists or of science in general. Rather, all of this is, from my perspective, nothing more than a recognition and description of a very ironic example of how the nature of experience, which includes the limitations inherent in the Individuals' creation of experience, makes unavoidable the presence of an experiential blind spot for each and every Individual, regardless of scale, and also regardless of profession, consisting of whatever experiences are the opposite of those which they are presently and actively creating and apprehending. The inability of scientists, as Individuals, to conceive of what the phenomena of wave-particle duality, quantum uncertainty, and quantum entanglement say about the nature of experience is ironic because *the very limitations of experience revealed by these phenomena are the same limitations that keep Individual scientists from understanding what these phenomena reveal about the nature of experience*. Thus, the revealed limitations are themselves concealed by the unavoidable functioning of the limitations that are being revealed.

It's a very sticky wicket indeed, and this sticky wicket, is exactly the same sticky wicket, the same set of experiential limitations, that are responsible for the functioning of what Vedantists refer to as maya, i.e., the situation whereby What's Actually There as Consciousness-Existence appears to Itself from the perspective of the Individual as the material, manifest, and phenomenal universe. That is, the same experiential limitations that hide from science what its own experiments reveal about the nature of experience, and so about the nature of all experiential reality, are the same experiential limitations that hide from us, as Individuals, both the True Nature of the universe as well as own True Nature as being ultimately composed of non-experiential Consciousness that, through relation to Itself, both creates and apprehends experience. Put another way, at a much more fundamental and subtle level of Existential self-relation and so experiential creation, the same experiential limitations that continue to pull the wool over the eyes of science, i.e., literally the I's of science, meaning Individual scientists, are the same experiential limitations that make it possible for Existence to pull the wool over its own I's, i.e., over Itself operating at the level of the Individual, and so hide from Itself its True Nature.

Thus, although it may seem that we live in a world of material cause and effect, we actually live in a world of Existential cause and experiential effect. That is, we live in a world where the cause is always some relation of Consciousness-Existence to Itself, and the effect is always the experience that is created and apprehended by the Individual Consciousness involved in that relation. However, the relations of Consciousness-Existence to Itself do more than just produce experience. That is, the effect of the relations of Existence to Itself have as their effect more than just the production of an experience.

If the relations of Consciousness-Existence to Itself produced only experience, then there would only be two complementary experiences that it would be possible for an Individual Consciousness to create and apprehend. That is, if the relations of Consciousness-Existence to Itself produced only experience and nothing else then those relations would only be able to produce, as an effect, the two most fundamental complementary experiences, i.e., wanted and unwanted emotion, because if the relations of Existence-Consciousness to Itself produced only experience and nothing else there would then be only two Existential relations possible; first level relations of aligned or oppositional Existential flow, producing for the Individual the experience of wanted or unwanted emotion, respectively.

However, the relations of Consciousness-Existence to Itself do not just produce experience as an effect. Rather, the relations of Consciousness-Existence to Itself also produce as an effect a Relational Structure that is composed of Consciousness-Existence as it is being in relation to Itself creating what it is apprehending as experience. And so the Cause produces an Effect and an effect. That is, the Cause, i.e., Consciousness-Existence, through relation to Itself, produces as a result or effect of any relation to Itself two different effects, one of which is composed of Itself, i.e., the Relational Structure, and the other of which is not composed of Itself, i.e., experience. And so the Cause creates Effect and effect, and the Effect, being not other than Cause, can once again serve as Cause and, through relation to Itself, create another Effect and effect, which Effect can serve again as Cause and iteratively on and on, ad infinitum, resulting in the creation of a fractal Reality Structure, a fractal Relational Structure, composed of Cause as it has become and is becoming progressively and iteratively structured in relation to Itself, while at the same time creating as effect a progressive series of experiential realities, extending from the emotional, to the mental, to the physical, that have as their basis the different possible relations of Consciousness-Existence to Itself made possible by the fact that the relations of Consciousness-Existence to itself produce not only effect, i.e., not only experience, but also Effect, i.e., Itself structured in relation to Itself as Relational Structure that then serves as the basis of a new Existential relation and so a newly created and apprehended experience.

Thus, the basis of the evolution of Reality and reality is not survival, because Existence cannot help but Exist. Rather, the actual basis of evolution, i.e., the evolution of Reality and reality as a whole, and not just the evolution of organic reality, the perceived evolution of which is just the tip of the evolutionary iceberg, is the desire of Existence to create and apprehend a new experience, a newly wanted experience. That is, Existence continues to project Itself into ever expanding levels of Self-relation and experiential creation because it wants to, and it wants to simply because it feels good to do so. In understanding the motivations of What Is Actually There in creating all of this, both as Relational Structure and experience, we need look no farther than our own motivations, as ultimately we are not other than That. Everything we do we do because we think that as the end result we will feel better, that we will experience a more wanted emotional experience. The rest of Existence is no different, because it Exists within the same parameters of experiential creation that we Exist, which is with the ability and necessity of choosing to create in each moment either a wanted or unwanted emotional experience as a result of choosing to be involved in a relation of aligned or oppositional Existential flow. Existence cannot help but Exist, and as it Exists it cannot help but be in relation to Itself and so cannot help but create, at the very least, a wanted or unwanted emotional experience. However, although each Individual point of Existence has no choice but to create some emotional experience, each Individual gets to choose the sort of emotional experience it creates, because each Individual gets to choose the aligned or oppositional nature of its fundamental and unavoidable relation to Itself. And since Existence has no choice but to choose to create one or the other of these opposite emotional experiences in each and every moment, it naturally chooses to create the wanted rather than the unwanted, it naturally chooses to create that which is attractive rather than that which is repulsive. That is the Nature of Existence and so that is our Nature as Individual points of Existence.

The difference between us, as Individual points of Existence involved for the moment in the Existential relations that create physical experience-reality, and the vast majority of Existence, is that most of Existence is cognizant of its role in the creation of experience and so consciously chooses its involvement in the fundamental and unavoidable Existential relation that determines whether it creates and apprehends wanted or unwanted experience, whereas we are mostly unaware of our role in the creation of what we apprehend as experience, in which case we are still choosing in each moment our involvement in the fundamental and unavoidable Existential relation, and so still choosing in each moment whether we create and apprehend wanted or unwanted experience, but rather than doing so consciously we are doing so unconsciously and reflexively. This is why we often end up creating the unwanted while trying to create the wanted, because without knowing it we are choosing to resist rather than allow, choosing to flow in opposition to our Self rather than in alignment with our Self, because in not understanding the nature of experience we must also fail to understand our role in the creation of experience. And in failing to understand our role in the creation of experience, experience is then seen as being Experiencer independent, existent as it is experienced to exist regardless of whether we are experiencing it or not. And owing to experiential entanglement, when experience is mistakenly conceived of as being Experiencer independent it then also mistakenly seems that the way to get to a wanted experience is by eliminating the unwanted and clinging to the wanted, when in actuality both of these attitudes actually unknowingly place us in relations of Existential opposition and so cause us to create and apprehend experiences that's have a quality of unwantedness rather than the desired wantedness.

Again, owing to experiential entanglement, when one concept is seen in reverse of its actual nature, any related opposite or complementary concept must also be seen as the reverse of its actual nature. And so when we conceive of experience as being Experiencer independent, which is not its actual nature, since its actual nature is that of being Experiencer dependent, we must then also conceive of how to create wanted experience in a way that is the opposite of the way it is actually created. So it is that we try to create wantedness through resistance, through self-opposition, and so we argue, we fight, we push against, we engage in wars, we try to eliminate the unwanted and cling to the wanted, attitudes known as aversion and attachment, respectively, because from within our inverted conceptual framework this appears to be the way to accomplish what is ultimately the prime directive of every point of Existence, which is to create and apprehend a more wanted experience. There is no evil, there is only Existence that's confused about how to go about creating wantedness.

And so we do not live in a material world, and so we do not live in a world of material cause and effect. Material reality does not cause Consciousness as an effect. We live in a world of Existential cause and experiential effect, where the relations of Consciousness-Existence to Itself are the cause and experience, which includes material-experiential reality, the effect. Therefore, the organic brain is not a material reality that produces as an effect Consciousness. Rather, Consciousness, through its relations to Itself, produces the Relational Structure composed of Itself that we apprehend as the organic brain. It is therefore not a question of how does the brain produce Consciousness, rather it is a question of how does Consciousness use the Relational Structure we apprehend as brain to create experience for Itself, to become involved in relations with Itself that create what it then apprehends as higher order physical experiences.

What we apprehend as brain is actually composed of Consciousness, as is everything, as is empty space. The same non-experiential thing that Exists directly where we each are as Individuals is the same non-experiential thing that Exists at every point in the universe and beyond. What Exists directly where you are is not your body, rather, what Exists directly where you are is the non-experiential Consciousness that apprehends the material experience of body. That what is there where you are appears to be a material body is no different than a reflection appearing to be what's there where there is actually a body of water. Thus, the ability to create experience, to apprehend experience, is intrinsic to every point in the universe and beyond. However, the type of experience created and apprehended is dependent upon the ability or way Existence can be in relation to Itself. And what the Relational Structure we apprehend as brain does is allow for Existential relations that would otherwise not be possible, and so allows for the creation of experiences that would otherwise not be possible.

For Consciousness to create and apprehend experience it has to be in relation to Itself and for it to create and apprehend a particular experience it has to be in a particular relation. The relations that create emotional experiences are different than the relations that create mental experiences, and the relations that creates mental experiences are different than the relations that create physical experiences. Consciousness cannot just decide that it is going to have a physical experience and produce for Itself such an experience in the absence of the Relational Framework composed Itself that allows for the particular Existential relation that produces as an effect that particular type of experience.

And underlying the experiential reality-reflection that we apprehend as the organic brain is the Relational Framework or Relational Structure composed of Consciousness-Existence that allows for the Existential relations that produce as their effect what Consciousness then apprehends as physical experience. And so again, the question is not how does the brain produce Consciousness, because it doesn't, rather the question is how does Consciousness, structured in relation to Itself in the way we apprehend as the brain, produce for Itself a particular physical experience?

But even more interesting is the question regarding how Consciousness, through its exercise of free will, through its intrinsic ability to choose its direction of flow relative to Itself, uses Itself structured as what we apprehend as brain to control Itself structured as what we apprehend as body. And it may be that this exercise of choice manifests in what is apprehended as quantum spin states.

Underlying every reflection is a reflective substance of some sort and underlying every experiential reality, every rock, every molecule, every atom, every quark, every gluon, every whatever, even space, is the Reflective Substance that is Consciousness structured in relation to Itself, Consciousness being in relation to Itself and as a result of those relations having configured and continuing to configure Itself into Relational Structures that are composed of Consciousness and so composed of, at each and every point regardless of scale, that which has the intrinsic ability to choose its direction of flow relative to Itself.

I used to think that quantum randomness was a function of the experiential limitations, a function of our complete inability to actually ever directly experience What's Actually There, because What's Actually There is ultimately non-experiential, ultimately of a Nature that is different or other than the nature of experience. Then I realized that there was a more simple and direct explanation, because underlying every experience, no matter what we call it, and no matter how small or large the experience, rests Consciousness that, like ourselves, is always free to choose to flow this way or that, in alignment with or opposition to Itself, according to how the Consciousness that is there directly is choosing to exercise its free will. And so the creation of any experience, which always involves some relation of Existence to Itself, always involves two choices, one of which we make as Individuals as we choose how to be in relation to What's Actually There, and the other of which is made by What's Actually There as it chooses how to be in relation to What's Actually Here, which in all cases involves Consciousness-Existence choosing how it will be in relation to Itself.

And since what we as Individuals create and apprehend as experience is the product of that relation, what we as Individuals create and apprehend as experience must then be the product of both of those choices, one of which we control completely and the other of which over which we have no control whatsoever, because both of those choices arise from and rest solely within the Consciousness that is Actually and Directly There, as a function of how the Individual Consciousness that is Actually and Directly There is choosing to exercise its free will. And because one of the determining factors in the creation of experience is inherently beyond our Individual control, the creation of experiential qualities other than those of wantedness and unwantedness must have some degree of unpredictability. The creation of the experiential qualities of wantedness and unwantedness is predictable because the other factor in the creation of experiential wantedness and unwantedness is the direction of flow of our More Fundamental Individuality, which is constant, and so the creation of experiential wantedness and unwantedness only varies as we, according to our exercise of free will, change our direction of flow relative to That.

You can offer numerous different Individuals the choice of ice cream or stepping off the side of a steep cliff, and no matter what it remains possible that one or more may choose the cliff rather than the ice cream, and you have no way of knowing which ones might do so or how many, because there is an inherent unpredictability in the Individual exercise of free will. An Individual will always choose what seems to create for Itself the most wanted experience, as that is its Nature, as that is the nature of Existence, but what seems to create the most wanted experience will vary with Individual perspective. And it is this inherent unpredictability in the Individual exercise of free will that lies at the root of quantum unpredictability, because experience is always the product of a relation, and in every relation there are two Individuals making a choice that determines how they will be involved in that relation, and it is the combination of those choices that determines what each Individual will, from their perspective within that relation, create and apprehend as experience.

Thus, from the perspective of the idealist the question is not why is quantum experience unpredictable, rather, the question is why should quantum experience be expected to be any more predictable than Individual behavior, since in both cases What's Actually There is Consciousness exercising free will? It only seems that experience should be predictable in the context of considering what's actually there to be consciousless matter, i.e., in the context of a materialist framework, in the context of a materialist conception of reality, where experience is seen as cause and Consciousness as effect. However, as has been shown throughout this work, in the opposite conceptual context, i.e., in the context wherein Consciousness is conceived as cause and experience as effect, the experiential effects, i.e., wave-particle duality, quantum uncertainty, quantum entanglement, and quantum unpredictability, rather than being paradoxical, become what is expected. Further, once these phenomena are recognized as limitations that arise naturally and unavoidably as a result of the way experience is created as the product of a relation in which the Individual that apprehends the experience must themself be involved, these phenomena, rather than appearing to be operant only in the creation of quantum experience, can be understood as manifestations at the quantum level of universally operant experiential limitations, i.e., experiential limitations that operate in the creation of every experience at every level, limiting what we can feel and know based upon what we are already choosing to feel and know, and dictating how we must feel and know based upon what we are already choosing to feel and know.

Article

Holographic Dreams

Iona Miller*

ABSTRACT

Our dreams, as well as our corporeal being may have roots in a holographic cosmos. This is reflected in the apparent holographic nature of our brains and the processing of underlying information encoded holographically within everything. Dreams are phenomenological. Jung thought they arise from the collective unconscious with an ability to transmit symbolic information to the waking psyche. We now know they help us incorporate new experience and consolidate long-term memory. They encourage brain plasticity and gene expression. Archetypes play a role in the unfolding of imagery and in the emergence of the Self and transcendent function, which can appear in an initiatory or healing capacity. Physical body symptoms, addictions, family and relationship problems, group conflicts and social tensions are mirrored in our night time dreams, and vice versa. All these experiences, even the most chaotic-seeming processes, when approached openly, reveal an inner order and coherence from the 'holographic blur' that provides new information vital for our personal and collective growth.

Key Words: dreams, REM, gene expression, psyche, Jung, hologram, process work, metaphors, initiation, psi, visions, holographic brain, energy fields, collective unconscious, fractals.

Introduction

The Field is the archetype of Process;
Cosmos is the ultimate symbol and archetype of Being;
Soul is the archetype of Life itself;
Soul transforms events into meaningful experience;
Internal phase coherence delivers the dream hologram;
The Self is Jung's archetype of wholeness, the God within us;
The central symbol of the Process is the creation of the world, its destruction and restoration.

The Unconscious is not separate from the body and the body is not separate from the Cosmos. When we turn our attention within we encounter the yawning Mouth of the Deep which contains the secrets of life and being. The unconscious is not separate from the body and timeless holographic history encoded in its wave genetics. The body is our deep memory, including our ancestral heritage. Highly charged psychophysical images encode somatic symptoms or dissociative disorders. We can guide ourselves to free negative residues of trauma frozen in the body and unconscious mind.

Processing dissociative reactions gently facilitates psychophysical energy release, leading to reintegration of dissociated and fragmented parts of the psyche. Therapy works best when it

*Correspondence: Iona Miller, Independent Researcher. Email: iona_m@yahoo.com Note: This work was completed in 2006 and updated in 2013.

emerges from one's inner healer, through one's own imagery, rather than via guided imagery or imported metaphors suggested from outside.

Epistemological metaphors -- how you know what you know and what it's like -- are gateways to the subconscious, as are dreams and symptoms. Content-free therapy can be done through metaphor, rather than through directly reliving trauma thereby avoiding re-traumatizing. In chaos theory, old forms must break down to make way for new ones to emerge. Healing occurs at the creative edge where new order emerges holistically. A small change in attitudes, the messages we send to ourselves, can make huge differences. Change the image and you change the associated attitudes and feelings.

Psyche Is Soul

What is the Psychè, the soul? Jung says Soul is the archetype of Life itself. Post-Jungian, James Hillman suggests that the soul is not a concept but a symbol. He describes it as a perspective, as deepening, noticing, penetrating, and insight.

Primal shamans were/are soul specialists, dream-walkers who understand its forms and destiny. Informational systems connect dreams and the body-mind. They claim the passions of the soul work magic and healing. Today's dream-healers serve a similar function characterized as soul recovery -- the reclamation of psychic energy.

Soul loss is the result of a split, dissociation, an abyss between our mundane lives and the vast depths of the psyche, in dreams and in nature. Symptoms include burn out, numbness, compulsivity, depression, restlessness, insomnia, and hyperactivity.

Healing is not necessarily synonymous with cure. Sometimes we heal physically, but not emotionally or spiritually. Sometimes we heal spirit and restore soul, but a physical cure remains elusive. In the past healing focused on the physical body and mind but not on the human energy field. When body, soul and spirit resonate, holistic integrity is restored.

We develop a relation with the sacred by accessing images and experiences in a dimension where magic and power reside, in which archetypes and entities have dreams, will, and intelligence of their own. Healing comes through visions, dreams, and symbols -- engaging the symbolic forms. Dream images have psychophysical correlates. The mind-body is a biohologram embedded in a holographic universe.

Images and dreams are direct paths to the unconscious. We call on the imaginal realm for healing, for wholeness. Like all living symbols soul resists all definitions, providing metaphors for primordial human thought. Hillman (1975) describes psyche's functions: (1) it makes all meaning possible; (2) it turns events into experiences; (3) it involves a deepening of experience; (4) is communicated in love; and (5) has a special relation with death. Vital essence is reintegrated through direct experience. It's an on-going process.

The Field (unconditioned consciousness, infinite information, and potential energy) is the archetype of Process. Jung called such pregnant fullness the Pleroma. Cosmos is the ultimate symbol and archetype of Being. Susskind describes the World Hologram (WH) in terms of information and levels of reality projected from the farthest edges of space. Information is fundamental and indestructible.

In this theory, the third dimension we experience is no more than a holographic projection of a 2-D surface. The event horizon acts as a hologram, and can actually describe any point in the Universe. That is, the physical information inside any cosmological domain should be holographic. There are several holographic theories, but that is beyond our scope here.

Symbols clothe the deep truths inaccessible to the conscious mind that can only be discovered through symbolic language, the vehicle the soul uses to express itself in dreams. Hillman adds that, "The soul is the component that makes possible the *meaning* and transforms events into *experience*".

Sleep and dreams harbor active experience-dependent processes related to neural plasticity. Ernest Rossi says, "We now know that sleep, dream, meditation, work, play, PTSD, dissociation, and other states of illness and health are associated with activity or experience-dependent patterns of gene expression and brain plasticity."

Dreams create experience-dependent changes, including gene-expression and proprioception. Rossi describes the creative replay of the "novelty-numinosum-neurogenesis effect". REM sleep elicits plasticity-related expression in previously activated neurons (Ribeiro). Internal phase coherence delivers the dream hologram. Fractality is charged self-similarity, which also informs DNA. Resolution depends on focus. With phase-lock, those dreams *might* come true.

All information of consciousness is carried, transformed and transmitted as holographic wave interference patterns on the surfaces of higher order hyperspace fields, which are resonant with the intermediate EM field of the brain. The entire universe, including all the visible and invisible structures within it, is essentially a hologram. According to the fundamental laws of electrodynamics, such information can be transmitted from one fractal-involved field to the other by phase conjugate adaptive resonance. (Maurer)

This experience *is* the 'Pyche'. Waking and latent memories reverberate in the absence of competing stimuli. Image takes the form of changing reality that surrounds us and lives in us, made of meaning and emotion that takes us back to an intimacy and a liminality otherwise inexpressible. It can be holographically reconstructed from neurologically transformed sensory images, or as stored memory field images.

Unconscious psyche and matter are one, unified as infinite holographic information. These inner and outer worlds remain invisible, inaccessible and inexpressible unless soul appears to guide us toward the *experience* of life. Archetypes are automorphic, subconscious holograms that express in dreams and waking consciousness. Nature always reflects the death/rebirth archetype of renewal and our sleep and waking patterns mimic her. Recognition and liberation are simultaneous.

We need only follow the divine song, the ancient sound, the Logos, the Word -- the audible life stream back to its undifferentiated, unconditioned source. We perceive a small portion of it as our individual stream of consciousness. Dreams bring the stream of consciousness into super-resolution, a superposition of past-present-future. One seemingly simple image in a dream can contain a full composite of the whole dilemma facing the dreamer at that stage in life. Our mythic body *is* embodied soul.

The dream has a central theme running through all its hyper-associative images, characters, and activities. Stories, including the ones we tell ourselves in dreams, are one of our most fundamental communication methods. Our brains feed on stories. Stories engage and activate not only language processing, but any other area in our brain that we use when experiencing the events of the story. Metaphors engage the sensory and motor cortex, and plant ideas, thoughts and emotions triggering neuro-chemical cascades.

The most obvious place to find archetypes is in stories. The righteous warrior, the wise-cracking sidekick, the villain who must be overcome, and the star-crossed love interest are all archetypes. Plot patterns can also be archetypal—the humble birth and prophesied journey of the hero, for example. Each archetype is referred to by the purpose it serves. In the archetypal world, everyone is the same, all around the world. Our emotional addictions to pain and suffering, contempt, insecurities, doubt, and failure is holographically-recorded and can be holographically healed. All archetypes are a form of human expression that is both holographic and physical.

We think in narratives all day long. We make up short stories in our heads for every action, contingency, and conversation. This activity continues in dreams through which a deeper part of our self conveys information. If the message is 'received' there is a sort of synchronous alignment of the conscious and subconscious energies. Jung claimed a psychological 'transcendent function' arises from a union of conscious and unconscious contents as well as the real and imaginary. Such coupling may occur during the dream, upon waking, or in therapeutic deepening.

Process work couples more brain areas in the communication, increasing the success of perception, transmission, and value representation. Metaphors help us relate to our existing experiences. We automatically link up metaphors and literal happenings with dream material. Dreams make us truly relate to the happenings of a story and activate parts of the brain that turn the story into our own idea and meaningful experience. Dreamwork couples action-based and perception-based processes. It is axiomatic in process work that every recall is a reframe. Information patterns are altered simply by accessing them.

The Voice of the Silence

Dreams have their origin in wholeness, manifest in polarity, and aim at totality. Dreams are the living voice and expression of Psyche, of soul -- a geyser of living creativity upwelling from the source of all. They are the channel between nature and culture -- between the unconscious and collective consciousness. We dream continuously as part of our stream of consciousness, the imaginal overlay and mythic interpretations we attribute to objective forms and events. This is the source of mythologizing and superstition.

[Jung] *decided that it was necessary to open his patient up to, not merely analysis of the unconscious, but to a true 'exchange' between unconscious energies and the patient's conscious ego. He asked himself: 'What do these dream figures wish us to do? What do they have to say about their condition and ours?' Of course, he was under no illusion that their advice would always be useful or of benefit to patients (he'd had a lot of experience with people suffering from schizophrenia), but he did feel it was important to let such 'personalities' have their say, also to allow them to enter into discussion with other unconscious figures, and even, on occasion, he thought it worthwhile to do their bidding in the real world.*

After four years scrupulously entertaining his own archetypal guests he realized that on occasion the presences had nothing to do with complexes associated with his own childhood. Similarly, the ones that seemed most useful for his own healing purposes (he was, remember, going through a period of intense introversion) were either mythological figures (he early on encountered his 'anima', for example) or, on some occasions, 'spirits of the dead' or spirits of socio-cultural warning (even prophesy). (Irvine)

In archaic times, people took it for granted that dreams were related to the world of supernatural beings in which they believed. They still heard the voices of their ancestors. Dreams served a special purpose: they predicted the future. There are many ways that dream symbols help us gain conscious self-knowledge. The unborn dream contains our hidden potential. Dreamers are known to interpret their dreams in terms of their own pre-existing beliefs, personal mythology, and worldview. Dreams can leave us in a mood, perplexed, or totally thunderstruck.

Shamanic Dreams / Archaic Realities

Ecstatic dreams, visions and trances allow shamans to maintain contact with their ancestors. Spirits of the dead communicate with the living in this way. When elders dream about the "immortals," they share the dream with the entire village. They reenact the dream with the elders playing the roles of the ancestors. Dream ceremonies help to align the present with the past, providing cultural continuity. Sometimes tribal members sing and dance each others' dreams, encouraging trust among tribal members.

Some communities hold Dream Circles, or morning dream-sharing sessions. Dreams are important because they are moments when we are stripped of rational thought. Retrospective analysis permits the full meaning of dreams to be found and new symbols to be created. In dreams we are in a spiritual state where "integral being" can emerge, connecting us with a deeper reality.

For example, some direct their dreams to someone who is several hundred miles distant; others can foretell both positive and negative events that will affect the community. People report precognitive dreams that issue warnings, describe a place they should not travel to or a person they should avoid, or future actions of a person toward the dreamer. Dreams also validate various aspects of the culture such as myths, songs, and social rankings.

The Gods live in the dark, hidden from the eyes of the people. Spirits give the initiate prophetic dreams, songs, and cures. Dreams carry him or her to the underworld to be instructed by deceased shamans. Many believe that they can ascend or travel to the heavens in their dreams, as well as to the "underground world." For shamans the power to dream is the power to participate in creation itself and dreaming reality is a sacred duty.

Some dreams are initiatory. During a shamanic dream initiation, the candidate usually experiences suffering, death, and resurrection, including a symbolic cutting up of the body, such as dismemberment or disembowelment by the ancestral or evil spirits. Shamanic power follows dreams of a dead mother, father, or other ancestors.

In his classic *Shamanism*, Mircea Eliade tells us that, "Ritual death can be achieved by extreme fatigue, tortures, fasting, etc. Seeing a spirit either in dream or awake is a certain sign of having attained some sort of spiritual condition, that is, one has transcended the profane condition of humanity. The chief function of the dead in granting of the shamanic powers is less a matter of taking "possession" of the subject than of helping him to become a "deadman", in short, of helping him to become a spirit too.

Sometimes initiatory dreams begin even in childhood. Such dreams of some future shamans may include a mystical journey to the "center of the world," to the seat of the "universal lord" and the "cosmic tree." The shaman makes the shell of his or her drum from the branches of this cosmic tree. If not heeded, the premonitory dreams of future shamans are followed by mortal illnesses. Conversely, a shamanic breakdown can precede renewal or rebirth as a healer.

Shamanic dreaming can be done wide awake, or in deep sleep. We each have a bit of the shaman within us. Shamans integrate their dreams into every major facet of their waking life. They don't recognize any rigid division between dream life and waking life. Loss of "contact with the sacred" results in loss of soul. Recovery requires initiation and successful integration of direct experience.

Anyone who dreams participates in shamanism, according to Dr. Stanley Krippner. He says in tribal life, "It is customary for dreams about the 'supernaturals' to be interpreted literally. It is typical for positive dream reports to be communicated only after their prophecy has been fulfilled. This retrospective analysis permits the verification of premonitions received in dreams and perpetuates, thereby, the use of dreams as forecasting devices. It also establishes the dreamer as a competent channel of communication with the spirit world."

Shamanic personalities work at the edge of chaos where it is often difficult to distinguish spiritual emergence from spiritual emergency, bloom or doom. Today's shamanism is connecting contemporary society with the mythic roots of humanity. Shamanism is beyond time; it's a primal spirit. Anything that is created is linked into that spirit.

Dreams are the interface between what exists now and what is coming into existence -- both earth and ether. We are all capable of transcendent awareness, of becoming shamans. The shaman is a shaman because he or she has been empowered by treading the road others wish to follow. The shaman is a symbol to others of their projection of a degree of personal insight and growth.

The shamanic principle appears in dreams, religion, healing, and transpersonal activities simply because its activity is essential to mindbody development. The inner shaman is an archetype which guides our exploration of self and dream worlds, transformation, and social flow.

The Unborn Dream

Personal limitations prevent us from experiencing unconditioned universal consciousness. In ordinary consciousness, our attention is limited to only a few of the nearly infinite variety of 'images' in the phenomenal field, which are theoretically available to individual personal consciousness. The phenomenon of resonance is implicated in this process. We resonate with those images which create and define our unique reality.

Dreams are ephemeral images, voices from other worlds. They may linger when we awaken in the morning, but often just as soon as we reach for them, they evaporate just beyond our reach. Our dreams are connected to those of others, to the gods and goddesses, to the Cosmos, to the Divine, and mysterious. We can only speculate on why we dream, but it seems to be related to working memory and regeneration. As we dream, we renew body and soul. Dreams call us to the heart of being. Elements from past experiences can become symbols.

Visionary experience emerges as countless images unfolding in rapid sequence because of the holographic nature of the process. So a single image may contain information about general attitudes, self-esteem, parental issues, childhood trauma, mortality, sexuality, and relationships, all at once. Each small part of the scene can contain an entire constellation of information. Any portion of a hologram or a dream contains the whole in lower resolution. So, they make valuable portals for entering the therapeutic process.

Dreams are the precursors to psychic growth and mindbody healing. Such meta-cognition can activate gene expression and brain plasticity. Mind-gene communication engages the transformational alchemy of mind, body, and spirit. Dream activities appear as metaphors for the dreamer's waking concerns. Many have claimed that scientific, technological, athletic, or artistic breakthroughs have resulted from dream recall.

Stan Grof has shown that the hidden holographic order surfaces nearly every time we experience a nonordinary state of consciousness. In nonordinary reality, such as dreams and shamanic journeys, momentary fissures in our constructed reality reveal a brief glimpse of the immense and unitary order underlying all of nature. During dreams, we can access the unending flow of wisdom, the flow of nature.

Krippner notes that some dreams are "anomalous" (or "psychic") because they seem to bypass the ordinary constraints of time and space. Examples include so-called telepathic, clairvoyant, and precognitive dreams that appear to incorporate another person's thoughts, activities occurring at a distance, or events that happen later yet appear to match the dream content. Persinger and Krippner showed that psi in dreams is strongest on nights that displayed the quietest geomagnetic activity compared to the days before and after.

About 17-38% report at least one precognitive dream which seemingly includes knowledge about the future which cannot be inferred from conventionally available information. Most studies indicate that women report more precognitive dreams than men, while the frequency of precognitive dreaming declines with age.

The occurrence of precognitive dreams correlates with dream recall frequency in general. Lange, et al say some experiences of precognitive dreams are illusions, i.e., coincidences between the contents of dreaming and waking experience. They get noticed due to frequency of dream recall and are given credence due to the combined effects of belief in the paranormal and a tolerance for ambiguity. Objective reality disintegrates into a holographic blur.

Holographic Blur

Our thought-forms are projected outward like a hologram in the theater of night. A hologram is remarkably similar to a lens in that it can make both virtual and real images. Life itself may be based on a holographic system consisting of coherence and interference. Order and patterns are the cornerstones of holography. Many scientists suggest the brain and body operate on holographic principles on the cellular, molecular, and neural levels.

Our brains can't tell real from vividly imagined experiences. If we imagine negative experiences the brain changes chemistry because it thinks that experience happened. This plastic capacity is the basis of the 'Re-frame' and 'Change History' in Neurolinguistic therapy. If you change the attitudes associated with events and memories it kindles a cascade of holistic changes in feelings and behavior. Such memory changes are consolidated in dreams.

The holographic model of the universe views matter as the constructive and destructive interference patterns created by interacting energy waves. Standing waves occur when a wavefront takes on a stationary appearance. As energy continues to pass through the system, each successive wave takes exactly the same position of the one before, making an illusion of stability. Conditioning prevents us from seeing the de-structured frequency domain. We filter out a portion of the information necessary to discern the true underlying pattern.

Holograms depend on standing waves for their existence. This is the wave nature of consciousness. The wave form profile of the self-similar fractal nature of life processes is continually generated at the quantum level. Our challenge is to bridge the mindbody gap to facilitate problem solving and healing that embrace all levels from quantum, molecule, genes, and brain to mind and conscious awareness.

In holographic theory, fragmentation created by boundaries does not exist. Each component is an embedded part of an unbroken whole. A problem with holographic theory is that we have little understanding of why some energy fields appear as stationary matter, while others are manifested as electromagnetic waves. Holograms are not necessarily created by light, but can be formed in the presence of any wave action. Creative sound is an acoustic superhologram.

To view the brain as a hologram, we must understand the mechanisms that create interference patterns. The holographic process involves both a reflection and reference beam. In the brain, past experience might serve as the reference beam. New incoming information is combined with the experiences (memories) of the past to create an interference pattern.

Almost immediately, new information becomes part of the "reference beam" and learning has occurred. When each new piece of information arrives at the brain, a new interference pattern is created and again becomes part of the reference background. A constantly shifting interference pattern provides the mind with a continually updated model of reality.

As other patterns from the environment mix with or beat against the internal rhythms, they will emerge to a greater or lesser degree as dominant energies and frequencies. These less frequent regularities of pattern will merge into concepts similar to the concept of self, but of slightly less importance. Because new sensory inputs that are similar will stimulate and activate these developing concepts, they will become the referents to which new stimuli are compared. Earlier layers of memory modify the later layers.

The complexes of energy fields remain active in the brain, forming a consistent pattern of self and memory with which new holograms are continually compared via holographic interference of waveforms. Consciousness thus develops a memory and a sense of self that is essentially the same 'self' that has existed over the length of time that the brain has existed.

As we think and perceive, the 'reference beam' which we refer to as consciousness interacts with the fluid hologram of experience and briefly attaches itself to a portion of it. The duration of 'nowness' is the present. We do not experience the 'now' as a single fixed hologram. Instead, we compare the present input with the totality of our past experience and the internal representation we have constructed of the immediate past. The interaction of sensory signals, our internal bodily functions, and our memory creates the total holographic interference pattern that we experience at any given moment.

Holograms create virtual images, three-dimensional extension in space that appears to exist, yet contains no substance. We generally believe that we are able to clearly distinguish between external and internal events. However, considerable research shows that the division is not as well-defined as we perceive. The "world-out-there" and the "world-in-here" are not always clearly delineated.

Carl Jung's theory of the collective unconscious is compatible with holographic theory. Jung observed that certain dreams, myths, hallucinations and religious symbols are shared by many people and cultures. According to Jung, these archetypes represent part of the collective unconscious derived from our two-million-year-old collective history. Archetypes are symbolically if not empirically real.

The holograms inside us are constantly morphing and adapting to us, as observers. The holographic material is alive and responsive, reflecting the thoughts, feelings, dreams and inspirations that we focus on. Finally, the inner holographic material that we engage with repeatedly is customizing our immediate personal reality and also setting-up the unfolding of future realities for us to experience.

Only a limited glimpse of the deeper order is available to us because we lack the knowledge to perceive or decode the frequency interference patterns. Dreams may be one way that we counteract our tendency to fragment the world. As shamanism has shown, dreams often reflect a hidden wisdom that exceeds our waking consciousness.

Dreams retain the psychic activity of the waking person, even in our sleep. Memory consolidation takes place in the quantum holographic realm of dreams. They present us with a seemingly chaotic jumble of imagery which is difficult to grasp with the rational mind. Our feeble attempts to comprehend them may or may not be viable interpretations. Furthermore, such explanations may evolve over time, producing new understanding and clarity.

Fractal Dream Domain

Entangled informational streams are organized around emotive charge into an interference pattern. At the greatest point of torsion we find the informational streams coalesce into an informational associative fractal dream domain centered around a 'strange attractor' linked to our personal issues. Jung called it a 'complex' -- a split-off part of consciousness.

The 'interference pattern' is the fractal dream domain, containing symbolic information that is the distillation of encrypted symbol knots of collective and personal information held as an energetic emotive charge with an associative fractal pattern of symbolic triggers and imagery.

Dreams exaggerate and distort but can disclose emerging physical change, including physical illness, which has escaped observation by daytime consciousness. We narrow our focus on the inner world of visions, hunches, etc. when we have to tune ourselves into the physical world. In that action we filter out much valuable input. When we fail to get the message dreams can repeat over and over.

In the past, there were dream temples that recognized the psychophysical connection to disease and healing. Individuals could incubate dreams in a safe and sacred place that mobilized spontaneous healing. The proper role of the rational mind in dreamhealing is to surrender to the autonomous flow of the stream of consciousness, and to suspend any analysis of dream material. The meaning of dreams is inherent in the experience. A temple healing was an epiphany with the healing god in the dream.

As humans evolved out of their more intuitive-instinctual relationship with nature and became more rationally oriented, dreams were interpreted in many ways. The phenomena always remain the same, but valid and reductive theories come and go. Extraordinary variations in the concept of dreams and in the impressions they produce on the dreamer make it difficult to formulate a coherent conception of them. The value and reliability of information processed as dreams has gone through as many changes as our culture, from shamanism to process work.

Holographic Model

The gist of the holographic theory is that our brains mathematically construct 'concrete' reality by interpreting frequencies from another dimension, a realm of meaningful, patterned primary reality that transcends time and space. The brain is a hologram, interpreting a holographic universe. And that includes the dreaming brain. The brain operates as a complex frequency analyzer.

Neurologist Karl Pribram hypothesized that the neurons, axions, and dendrites of the brain create wave-like patterns that cause an interference pattern. Pribram explained many of the mysteries of the brain, including its enormous capacity for storage and retrieval of information. Memories are not localized in any specific brain cells, but distributed throughout the whole brain. In dreams we spontaneously construct a parallel reality that may include telepathic dreams, lucid dreams, and transpersonal dreams.

Pribram postulated a neural hologram, made by the interaction of waves in the cortex, which in turn is based on a hologram of much shorter wavelengths formed by the wave interactions on the sub-atomic level. Thus, we have a hologram within a hologram, and the interrelatedness of the two somehow gives rise to the sensory images. There is a more fundamental reality which is an invisible flux that is not comprised of parts, but a holistic inseparable interconnectedness -- the primordial source field.

Since thoughts are three dimensional, then dreams are holograms. Yet dreams are holograms of information, conversations between different facets of our awareness, intent on communicating, remembering, and healing. They encode and consolidate our emotional memories, anticipate our futures, and review our past.

Our dreams are holograms which are three dimensional visual pictures. Usually, in a dream, you occupy a body much like your own physical body, but it is a virtual body in a virtual space. The dreamer remains in the position of perceiver analyst even in the altered state of awareness. The dream is a vision loop that sometimes repeats its message over time. Sometimes it confides its secrets to our conscious awareness, provides insight, or even innovation. Thus, consciousness dreams itself into existence.

A dream is the experience of images, sounds/voices, words, thoughts or sensations. During sleep the dreamer is usually unable to influence the experience. The scientific discipline of dream research is oneirology. Dreaming has been associated with rapid eye movement (REM) sleep. REM is a lighter form of sleep that occurs during the latter portion of the sleep cycle, characterized by rapid horizontal eye movements, stimulation of the pons, increased respiratory and heart rate, and temporary paralysis of the body.

Phenomena associated with the brain can be explained with the holographic model. They include the ability to remember and forget, to recognize familiar things, photographic memory, associative memory, the transference of learned skills, the vastness of our memory, etc. Our brains are truly and completely of a holographic nature.

The holographic model explains many general phenomena not yet fully understood by conventional science. They include out of body experiences, near death experiences, multiple personality disorder, dreams, lucid dreams, placebo effect, psychokinesis, auras, and human energy fields.

Fred Alan Wolf believes dreams are actually visits to parallel realities, and the holographic model will lead toward a "physics of consciousness". The physicist says that universe itself may be a giant hologram, quite literally a kind of image or construct created, at least in part, by the human mind. According to Wolf, in dreams our consciousness moves to parallel levels of reality.

Wolf offers a quantum mechanical model for the emergence of self awareness from holographically-generated dream images. He claims, "Self-awareness arises from the ability of a simple memory device, an automaton in the brain, to obtain images of holographically-stored glial cell memories and, most importantly, through a quantum mechanical process, to also obtain images of itself."

He says, "*Each self-image is composed of a quantum-physical superposition of primary glial cell images and an image of the automaton containing those images. These self-reflective images are ordered according to a hierarchy based on increasing levels of self-inquiry conducted during the dreaming process*."

Wolf's idea is based in part on the work of Renato Nobili, which shows that ionic wave movement is similar in form and structure to quantum waves but different in certain essential details. It occurs in the glial cells of the brain making it an ideal medium for supporting and producing holographic imagery.

"Nobili (1985) proposed that Na+ and K+ transport through glial cells in the form of oscillating currents producing wave patterns--in which the motion of the sodium ions effects the movement of the potassium ions and vice versa--that satisfy a Schrödinger Wave equation. He found that contrary to lightwave holography, Schrödinger wave holography was far more efficient in producing holograms in glial tissue. "

"He also discovered that the close proximity of signal sources and receptors (which is in itself in good agreement with other neuro-physiological cortical diagrams) in the cortex was ideal for both production of reference waves and information wave recovery. Complete details for the mechanism for glial imaging/holography are in his paper," (Wolf).

Bokkon has done extensive research on dreams.

...Electrical signals generate visible pictures if electrical signals are converted to electromagnetic waves (EMW) of the visible range (light photons of wavelengths between 350-700 nm). During dreams our eyes are closed, so the brain is isolated from visible EMWs of the surroundings, yet we can see visible dream pictures. It follows from the foregoing that electrical signals of the brain processes can generate visible pictures of dreams if and only if electrical signals are converted to weak, EMWs of the visible range (biophotons) in the brain. The permanent, ultra-weak electromagnetic light photon emission from living systems is called biophoton emission. However, nobody can explain origin of visible dream pictures if only the laws of physics are being taken into consideration." p62.

Bókkon notes that thought and light (waves and biophotons) act as carriers of information and action. He postulates that light can be a carrier of consciousness. Most of the NDE accounts will substantiate this. In the NDE accounts, people and God are described in terms of light. He also discusses that sounds, smells, tastes and touch are all converted to electric signals and then to biophoton signals in the brain. Biophoton processes can explain synesthesia, another common NDE phenomenon. p62.

The way that the brain processes information is much like a holograph by "biophotons generating visible (conscious) and invisible (unconscious, metaphor-like) pictures in the brain." Pp 62-63. There are several precursor conditions for a holograph to work in the human body. One observation is that living cells produce and use electric, magnetic, electromagnetic and acoustical waves interchangeably. Living cells are capable of producing coherent and incoherent photon waves. Cells can be tuned to a permanent signal-bearing frequency.

Another relevant bodily process is provided by non-calcite piezo-biocrystals. "Piezo-biocrystals can work as pieces of information storage units by biophotons in a holograph like way in the brain. This idea was based on other kinds of biocrystals found in the living cells." p63. Bókkon states that biophotons explain the association of different information processes in the brain. "When two coherent light waves or an incoherent and a coherent wave are penetrating into a biocrystal, light waves can exchange their information in an associational way." Id.

One of the most important issues in memory retrieval has to do with temperature regulation. Not surprisingly, "temperature regulation is important for many holographic systems which use different electromagnetic photons, because holographic pictures can be deformed if temperature fluctuations are too large." p.63. Temperature also assists in developing optimal explicit memory through a structured neuron system. This is necessary for "strong synchronization of different electric and biophoton signals and increase the signal/noise ratio." p63.

Bókkon discusses the biological and chemical reactions that occur during NREM and REM phases of sleep. During NREM, *"there is a decrease in blood pressure, heart rate and respiratory rate, while REM sleep is associated with activation and increase in the irregularity of these functions." p64. Although neurotransmitters play a role in the neuromolecular processes, they also play a role thermoregulation and in sleep/wake cycles. Id. The observations about neurotransmitters and synapses from the sleep process do not address information exchange in the biophotons. Bókkon gives the example: "If depressed patients take antidepressants or sedative drugs they can calm down. However, drugs change neurotransmitters levels and many physiological functions (thermoregulation) in the brain, but long-term information does not change in their brain." p66. He makes the point that there is a distinction between brain molecular processes and memory.*

Bókkon postulates that it is *"probable that informational processes are connected to biophysical (biophoton or quantum) processes in the brain. Biophoton processes arise from molecular (thermoregulation) processes in a cooperative way, and they influence each other." p68. He notes it is impossible to "perceive visible pictures without visible EMW/biophontons during dreams. Dreams are not dependent on the presence of contemporaneous visual-perceptual experience. So, there must exist an informational system of biophotons in the brain." Id. "Dream pictures make a connection possible between the explicit and implicit memory systems in the brain, which allows us to remember our unprocessed emotional information. But, visible or invisible dreams are metaphor-like, because dreams are nonlinear holographic informational processes, and several visible/invisible pictures are assembled together. . . The possible origin of dream emotionality is REM-associated limbic activation, because all input information, which goes into the brain also goes through the limbic system" p68.*

The REM state is different than dreaming. *"The REM state is a cholinergic mechanism that is motivationally neutral, but the dream states are of dopaminergic mechanisms. L-dopa causes an increase in the frequency and intensity of dreaming without any effect on the frequency of REM sleep. p70. "Dream imagery is not generated by chaotic activation of the forebrain, but rather by a specific forebrain mechanism with dreams and complex cognitive processes." Id.* This is the mechanism for lucid dreaming. *"REM dreams are especially important, because they allow a connection between the explicit and implicit informational systems of the brain. However, we remember our dreams if information of dream pictures can go into our consciousness." Id. Interestingly, drugs can inhibit the replay of associative visible pictures in the forebrain. But dreams, as informational processes, work continuously in spite of whether we can remember them or there is a lack of REM during sleep. p71. Another interesting fact is that the brain can use dynamical pictures via biophotons, and therefore, the hypnotic state allows a direct connection with implicit information of the brain which works by pictures. p72. Hypnosis and dream analysis have been used with a lot of success in behavior modification.*

In the footnote, is an interesting comment on the phase type hologram. Bókkon writes, *"The phase type hologram produces phase changes in the reconstruction beam due to a variation in the refractive index or thickness of the medium. Phase holograms have the advantage over amplitude holograms of no energy dissipation within the hologram medium and a higher diffraction efficiency." I have always wondered if the transition from life to the other side involved a phase transition of consciousness. This makes sense when one considers descriptions of NDErs regarding the fourth (or greater) dimensions. Is it possible to key into the biophoton process to enter other dimensions with our consciousness, such as the bi-location, inter-dimensional and time travel*? (Bokkon)

Jeanette Mageo (2004) suggests,

Contemporary psychologists hold that dreams sort memories. These memories can be seen as shared memories symbolizing key contradictions in culture. These contradictions originate in psychologically unresolved historical problems that rupture shared meaning systems. Dreams evoke these contradictions holograpically by deploying images circulating in the public sphere that constitute meaning fragments. Dream images are often dramatically fragmentary, which compels dreamers to elaborate them. These elaborations constitute cultural work: Through a figurative mode of thought, dreamers connect a fragmentary image to personal emotions and to other significances in a culture's symbolic–psychological world, thereby making new meanings. Dreams and interpretative work on them, therefore, are usefully understood as instrumental to cultural change and as instances of cultural practice rather than solely as private and individual. The article develops this holographic theory of dreams through the analysis of 2 Samoan dreams from the 1980s and 1 American dream from the 1960s. (Mageo)

Dream Healing

There are several systems for accessing a feeling-identification with dream images, but they rarely lead to whole healing of the psyche and body. Some people feel they really "get" a dream when they experience the moment of "a-ha" or integration. The problem here is that stops the process of relating to the dream image by substituting some sort of intellectual inner "click" which may or may not be "right."

Dreams have many levels of reality which no single interpretation can encompass. A myriad of interpretations contain useful self-knowledge. Even a single dream can continue to unfold over the years since it contains an unfathomable depth of information. In process work we evoke deeper images of the same fractal to de-structure consciousness down to more primitive forms and levels that could be called holographic healing from the zero point.

Beyond the symbols, beyond the "click," beyond "a-ha" is a healing state. It is a gift from your dream in the form of a healing state--a place which is without dialogue, which is about vision, which is about healing inside, and which is beyond mere psychological understanding. This is Mystery.

So much of our time and energy is invested in building up models allowing us to formulate our ego view of the world of relationships and preferences. Where the most profound healing comes in is in the holistic (body/mind) experience of the dream. When you re-enter a dream in therapy, both the conscious mind and subconscious cooperate in a new and wonderful way that you may never have experienced before.

Unpleasant seeming dream imagery often transforms into a peaceful, healing place, if you allow the imagery to take your consciousness down into deeper, less structured awareness. The healing comes from simply "being there." This is a far cry from the scientific or heuristic understanding of dreams. However, Freud was not wrong when he postulated that the dream was the result of the conflict or cooperation of psychic forces.

The process that underlies dreams, when studied, can reveal the nature of these psychic forces. One of the main focuses in modern dream healing is on actualizing the healing power within dreams and other visionary consciousness states. There are many things you can do with a dream. In lucid dreaming you become conscious within the dream and direct your activities as in waking life. This may produce an increased sense of personal power and control. However, there is a chance that this is an invasive intrusion on natural corrective forces by an over-active ego.

The point of dream work is not to take the ego into the dream world. We need to bring the dream images into our conscious awareness and waking life. Since the dream state arises from beyond the ego, anything can happen, and natural laws of physical reality do not apply. Unbounded by any physical limits and laws, dream realities broaden awareness so that we can begin to experience our full range of humanness.

Virtually anything is possible in the dream reality -- death, rebirth, time travel, out-of-body journeys, enhanced physical or mental powers, even extraordinary effects like healing and balancing. Yet, there is a voice in most of us that wants to discount the dream experience as a less important, inconsequential reality than our waking experience. For example, a parent tells a child, "Go back to sleep; it was only a dream!", after the child has just awakened from a terrifying dream and still experiences the physiological consequences, which are very real.

Disregarding the nightmare is one way to ignore the power of the dream as if it did not have impact or validity in the conscious awareness and experience of the child. The truth is that the experience of a nightmare is just as threatening and dreadful as any waking situation that evokes extreme fear and bodily contraction. In fact, the nightmare may usher in an even stiffer fright because it may be drawing on the fantastic and other-worldly aspects of the psyche.

What is important to observe is that, in both cases, the fear experience causes bodily feelings and reactions. Our natural reaction to a fearful situation like a nightmare is to turn away and avoid the experience entirely. This avoidance (a version of "out of sight, out of mind") sets our system off-balance and triggers the fight/flight syndrome. To re-establish the balance and harmony, it is usually necessary to stop avoiding the fear and turn around and move toward it, accepting it and owning it as a valid part of our reality.

The monsters of our dreams are only alienated parts of our self, vying for attention. If we can embrace the fear, we no longer need to run away, and we can experience the peace that comes from having "let go" of the fear. Pain, either physical or emotional, is a marker that indicates where healing is needed within us; but we usually surround our pain with fear to protect us from experiencing it. The fear is usually a base for our anger, or any of the other numerous denial and avoidance strategies we use.

The nightmare makes us a gift of the fear and its underlying pain. It leads us to the inner places that need healing, and provides the healing as we experience expansion within of our "stuck", blocked, lifeless parts. At the heart of our approach is the notion that because dreams affect us on our primary experience level -- the body -- and can stir intense multi-sensual feelings and reactions in us, dreams can be used to enter a bodily place of dis-ease and restore the natural flow and balance to that place. In honoring the dream we draw from the ancient healing tradition of the past, and the best of modern psychotherapeutic technique.

The ancient word for therapy, *therapeuin*, originally meant "service to the gods." In this case therapy facilitates the healing process of the Greek god Asklepios. He was god of both dreams and healing. The content of the dreams -- the characters, the inanimate objects, the activities, the feelings, the colors -- can all be doorways into the infinite inner territory of our myriad inner selves. They are states of consciousness that facilitate healing on mental, physical and spiritual levels. If we can go deeply into the experience of a dream such as the nightmare, for example, we can bring a healing to the dis-ease that caused the nightmare.

Dreams and nightmares are a unique way to move our awareness into our inner feelings and bodily places of flow and blockage. With a remembered dream, we already have in our grasp a good start at an inner resolution of the process. Borrowing from C.G. Jung, we propose the idea that dream symbols arise from the psychic energies that create us and bind us together with all other life forces, the collective unconscious.

However, moving beyond analytical and interpretive methods of treating dreams, it is possible for us to experience directly the timeless and dimensionless primal force that creates dreams. To do so we have to use dreamhealing to travel beyond the symbols to their very source. We call these experiences dream journeys, in the old shamanic sense. The therapist functions as a guide to take the client deeper than the surface symbolism.

Symbols are merely a means of capturing our attention -- of attracting, appeasing, or scarring our ego's conscious waking awareness. Any illness or disease, as the name itself suggests, has at its source a state of dis-ease or out-of-balance energies. Like the shamans of old, Jung noted that the onset of any serious disease was reflected in dreamlife.

In addition to leading to the source of our dis-ease, dreams and nightmares also have within them the potential for expansive experiences which can heal and bring us back to a state of balance and health. They are both diagnostic and prescriptive, in that sense. They reveal both problem and solution, if we only learn how to attend to their clarion call. On the surface and analytical level, dream symbols usually relate to the ego's particular concerns.

Some "big dreams" carry a more mysterious, archetypal or collective value. However, each symbol is actually of equal value. They are doorways opening into the formless, chaotic energy underneath it which gave rise to it. Interpreting the symbol gives us a more detailed description and picture of the doorway, but does not give us the experience of going beyond that doorway and exploring experientially what is on the other side over the threshold in those primal energies.

Dream healing centers around the idea that by going into and then past the experience of the symbols, we can experience the consciousness that created them. This creative state is a source of healing and re-creation. Some symbols offer access to memories of the past, some reveal future events, others can lead us to our inner healer -- the part of us that can provide the energy we need to restore balance and harmony within ourselves.

Much work has been done with imagery and healing, usually importing symbols or images into the client's visualization. The healing tale or teaching tale is used in both spiritual and secular counseling. The "imported metaphor" is part of the stock-in-trade repertoire of Ericksonian hypnosis. The results are inherently stronger when the individual produces their own imagery while the therapist unobtrusively helps the client avoid the pitfalls of self-indulgent fantasy.

The client is guided to stick with the metaphors that arise from within to describe what his state of being and experience is like. You can experiment with this yourself, simply by asking yourself a few simple questions: What would you like to have happen? When it isn't happening, how do you know its not? And where do you feel that in your body? And what's it **like?**

By this means you create your own metaphor for your personal experience, whether it comes from dreamlife or some life problem, or a childhood trauma. The therapist functions solely as a guide to the inner realms, since it is familiar territory to the practitioner. We can use the well-known map analogy, noting that the map can only be a partial representation or symbol of the actual terrain.

For example, looking at any map of the countryside we can see lines that mark rivers, hills, and other topological features; however, to walk through an actual old growth forest with a compass, climb the hills and pitch camp under the protective canopy of the trees, and listen to one of those rivers imprints a much deeper impression of the forest than the map ever could. It is a full experience of what is behind the map.

Trying to experience the terrain through the map is like interpreting the symbol, while the experience of going into and beyond the symbols is as ever-changing and alive as an excursion deep into the forest and the mysteries of nature. Another example of the distinction is the difference between reading a recipe and tasting the dish. The savor certainly isn't the same.

A dream guide, like a river guide, takes the person through the turbulent (chaotic) waters of the psyche, past the rocks and boulders of their fear, to find the safe passage where the river flows easily into the calm beyond the rapids. The therapist's approach evolves in the moment to keep pace with the flow of the client's process deep in the heart of the dream. Consequently, the client has an active part in the healing process and learns psychological self-care. Flowing with the experience through the progression of multi-sensory images provides the pathway to healing.

The experience of finding an inner healing state is invaluable, as it teaches firsthand that the healer is within. The outer healers are only representations or mirrors of what is already inside. The healing process and myth are deeply engrained in our lives, as individuals and societies.

Each culture evolves its own variations on health and disease, and those able to aid in recovery from physical and mental distress. The problem with the old western healing paradigm is that the perception is that healing comes from without. In our culture now we are developing many alternatives to mechanistic medical and psychological practices. One of those alternatives is awareness of personal mythology.

Jung suggested that each individual life is based on a particular myth. By discovering that myth, we can live it consciously and adapt ourselves to our destiny, thus harmonizing inner and outer experience, and allowing our true individuality to emerge. But mythic living doesn't necessarily mean living one myth, since the patterns of all god/dess forms are within us.

The myth does not provide us with a blueprint for daily living concerning what we should or ought to do. Instead, it helps us in the process of discovering who we are, where we come from, and where we are going. They spark our sense of discovery and urge us to question and go deeper. There is a chaotic assortment of mythical images within each of us, but sometimes certain themes emerge and assert their priority on a life. So, an individual life seems to strongly parallel a specific myth theme.

One way this can manifest is through an uncanny series of synchronistic events, wherein a particular myth becomes the paradigmatic model of a life. The quest for actualization of this myth motivates a variation on the age-old journey of the hero. In ancient Greece, if you wanted to ensure success in some undertaking you invoked the god who oversaw your particular endeavor with prayers and offerings.

As stated before, Asklepios was the Greek god of healing and dream. He was the son of Apollo and the mortal Coronis who was slain by Apollo for infidelity before the child was born. Taken prematurely from his mother, Asklepios was raised by the centaur Chiron, who was a master of the healing arts. Asklepios was an able student who soon surpassed his teacher and incurred both the wrath and blessings of the various gods and goddesses. To protect him from these whims, Zeus immortalized him as the Divine Healer. An entire healing tradition developed in ancient Greece based on this myth.

The medical physicians became known as Asklepiads; however, the Asklepian dream healing temple was the place to go if their medicines and treatments failed. At the Asklepian temple, the god himself visited mortals in their dreams to bring divine healing. At the temple, Asklepian priests oversaw the rites and procedures which brought the sick mortal into direct contact with the god. These temples were located at great distances from the cities and populated areas in Greece so that to reach one a pilgrimage was necessary.

Having arrived at the temple, one was received by the temple priests who began the sometimes lengthy process of determining whether or not the god had summoned one for healing. The priests did have a therapeutic function in the temple, but they were not in any way therapists, nor did they interpret any of the supplicant's dreams. The priests determined whether or not one had been summoned by Asklepios by making discreet inquiries about the god's appearance in their dream life.

An appearance by the god signified that one had indeed been invited and was ready to enter the temple. The form which Asklepios assumed in dreams was either a snake, or less commonly a dog (or wolf). The next steps of bodily and mind purification were begun. Another interview with a priest was held because it was recognized that unless a person was conscious and accepting of his present life condition he could not expect a healing from the god. After the interview, the patient's body went through cleansing in the springs or streams around which the temples were always constructed. And at last, the supplicant was prepared to approach the god.

Since Asklepios visited the sick in their dreams, a special chamber, called the abaton (a Greek word meaning "a place not to be entered into uninvited"), was provided where the person would remain alone and asleep. The couch inside where the patient lay was known as the kline. This period of waiting for the god was called the incubation. After dreaming the patient was interviewed by the priests who, without interpreting the dream, would instruct the patient as to whether or not the god had brought the healing.

Sometimes many sessions in the abaton on the kline were necessary to come into contact with the god and the sick did not leave the temple until they were healed. As far as interpreting the dream, the belief was that the experience of the dream and not an interpretation was how the healing came to the sick. The healing was accomplished through the direct intervention of the god himself with the patient's soul through the dream.

As a final part of the healing process, a fee was paid to the temple priests as an offering for the ongoing maintenance and work of the Temple. It was said that a failure to do so would result in a relapse of the dis-eased condition. Testimonies were inscribed on the temple walls attesting to the miraculous and powerful healing which went on in the temples, including cures for afflictions like blindness.

In this way the Asklepian dreamhealing went on for hundreds of years. This tradition is continued in dreamhealing. The eight phases of dreamhealing reflect an archetypal healing process. This healing myth is reiterated in the techniques ("ceremonies") of many disciplines.

These steps form the real sequence of inner healing no matter what the outer form, including traditional medical practice. These phases include: 1) the pilgrimage; 2) the confession; 3) purification; 4) the offering; 5) dream quest; 6) dreamhealing; 7) work on dreams; 8) re-entry or integration. The entire process is contingent on a healing sanctuary, whether that refuge can be found without or within. Processing is the principle of assisting an individual to look at his own existence, and improve his ability to confront what he is and where he is for greater adaptability, wholeness, and health.

Arising of Images in Dreams

Ernest Rossi suggests, “new experience is encoded by means of protein synthesis in brain tissue...dreaming is a process of psychophysiological growth that involves the synthesis or modification of protein structures in the brain that serve as the organic basis for new developments in the personality...new proteins are synthesized in some brain structures associated with REM dream sleep.”

Rossi describes the four creative stages of the processing of self-reflection in dreams. During an ordinary dream, processing takes place at the lowest levels of image production wherein no sense of self is present, i.e., the dreamer is unconscious of having a dream.

This corresponds to the formations of non-self-reflecting states, level (0), where the dreamer is not aware of being present in the dream and the dream is experienced unconsciously with little sense of identification of self and other. Images are seen, but not self-reflectively experienced since no self has been defined. This constitutes unconscious data processing. Images are there but they have no meaning.

At level (1) the dreamer becomes involved in the dream. Moments of conscious and unconscious dream experience result. This corresponds to jumps between level (0) and level (1) and the formation of images mostly in the lowest level. When level (1) is reached, emotions are felt during the dream, but as soon as the dreamer ascended levels, only images are present. By jumping between levels (1) and (0) various potential emotions are aroused and sensed consciously during the dream.

Meaning arises when the self appears. Such events appear as enlightened or intense awareness. When level one is achieved the dreamer not only sees the images, she or he also possesses feelings about them that can be associated within space and time boundaries. These are primarily body images emotionally felt. Perhaps this is a clue to illness arising from emotional causes probably initiated during early childhood. (Wolf)

At level (2) the dreamer has thoughts during the dream. Thoughtforms arise and jumps between levels (0), (1), and (2) occur, resulting in a range of states of conscious and unconscious thinking, feeling, and observing. Thought adds much greater meaning to the emotions and the observations. Just as images come from sensory inputs, thoughts and feelings are capable of outputting in terms of expression of words and feelings. Unconscious thought forms integrate or superimpose conscious emotions and thus tend to have no emotional content per se.

However, as thoughts are expressed in words, i.e., self-reflection occurs, emotions can and often do arise, sometimes unexpectedly. This would be due to a jumping from level (2) to level (1) as a result of inquiry. Most likely the jump downward in the hierarchy involves a disruption of the higher level functioning. Thus, during the wake state, when words move us to emotional action we are descending to a "lower" level of self-awareness.

At level (3) the dreamer is able to simultaneously be aware of the previous levels of participation and observation during the dream. Here the sense of self more fully emerges as the dreamer deals with archetypal images. Again there is full access to the lower levels.

At level (4) the dreamer consciously reflects on the fact that she/he is dreaming. This would be the super-archetype state corresponding to lucid awareness during the dream. At each level the automaton is able to record images from lower (higher on the Figure) levels or images of the highest level it is capable. The tendency is to descend levels more readily than to ascend them.

Intuition suggests that these levels are somewhat akin to energy levels of atomic systems, and concepts involving entropy may be applicable. Descent results in less self-awareness (greater entropy) and therefore a requirement for more automaton-mechanical behavior. Ascent results in greater choices, to become aware of existence in other "worlds". More complex imagery with a higher number of paradoxical simultaneous features becomes knowable.

It is only through passive awareness within oneself that simultaneous knowledge of all of the images is possible. Also remember that any attempt in asking about one's state of awareness and communicating this to the outside world in this model alters the state in correspondence with the uncertainty principle.

This could have profound effect on dream research where the researcher attempts to communicate with the dreamer while a dream is in progress, perturbing it. Of course there would be more levels requiring the formation of more complex images, and greater mind-ability. At the highest level, we reach a state of "pure" awareness with surprisingly no images present.

Discussion: The Origin of Dreams

There are many competing theories for the neurological cause of the dreaming experience. REM sleep is commonly associated with dreams, though it is not known whether dreams actually occur more frequently during this light sleep stage or are simply recalled more easily. REM sleep is known to be produced by a brain region known as the pons. Dream sleep and dreaming are associated with high levels of brainstem arousal and right cerebral activity.

Dreams are generated deep in the back part of the brain that processes emotions and visual memories. The fact that REM sleep continues normally is significant, because dreaming and REM sleep occur together, though research has pointed to different brain systems underlying the two. Research appears to confirm that dreaming and REM sleep are driven by independent brain systems. The origin of dreams is a process of memory. Memories are stored in the mind by the brain hippocampus involved in the formation of dreams.

As further evidence of dream's chaotic nature, a case can be made that dreams are holograms. In using dream symbols for healing work, it seems that "all roads lead to Rome." A person can enter the dream through any of the symbolic doorways and derive a very different experience of consciousness along the way. Still, following the symbols deeper and deeper one arrives at that healing, primal level.

As long as the image is followed back faithfully, the connection can be made from any beginning point in any dream, old or new. That is one reason we never need to go back to a particular dream symbol that has not been worked, but can pick up the process entering through a fresh dream image. The deeper mind always presents the best departure point for current conflicts and turmoil, that which seeks healing.

Conclusions

Waves Carry the Dream. The part is contained within the whole, but the whole is also contained in each part, according to the holographic model of reality. Therefore, by changing the part, therapy changes the whole. So philosophically, it makes sense to approach the whole person, rather than the part which contains less-detailed information. If the therapist chooses the part to work on, it is limited.

Psyche has many options to choose from its deeper wisdom. Small changes in therapy can translate into exponential changes over time, since chaotic systems are sensitive to initial conditions. They quickly pump up small changes to larger changes in awareness. Part of the overall healing model is that WE ARE NOT SEPARATE. And, to the extent that we can embrace that model, it becomes more of a reality in the most profound sense.

In this dynamic model, there are no "things" only energetic events. The holoflux includes the ultimately flowing nature of what is, and also of that which forms therein. Bohm speaks of "the source" as beyond both implicate (enfolded) and explicate (unfolded) realms. We can imagine Source as the coherent Light which illuminates and objectifies the implicate realm.

Remember, the psyche unfolds just what each participant needs for healing in a nonrational way we could never guess or make up. This process is initiated when we deliberately "observe" the stream of consciousness in a therapeutic context. Dreams are such raw experiences and imagery. The importance lies in the experience; meaning is inherent within it, embodied as a gestalt. We have simply observed that shamanic consciousness journeys and holographic theory are analogous in many ways and may shed light on one another.

References

Achterberg, J. (2002). Imagery in healing: Shamanism and modern medicine. Boston:Shambhala.

Bohm, David, 1980, *Wholeness and the Implicate Order*, London: Routledge and Kegan Paul.

Bókkon I. (2006) Dream pictures, neuroholography and the laws of physics. *Journal of Sleep Research*. Vol. 15, Supplement I. Abstract. p:187.

Bókkon I. (2005) Dreams and Neuroholography: An interdisciplinary interpretation of development of homeotherm state in evolution. *Sleep and Hypnosis* 7, 61-76.

Eliade, Mircea, (1951), Shamanism, Princeton University Press.

Grof, Stan, (1993), *The Holotropic Mind: The Three Levels of Human Consciousness and How They Shape Our Lives*, HarperOne; Reprint edition (May 28, 1993).

Ingerman, S. (1991). Soul retrieval: Mending the fragmented self. San Francisco, CA: HarperCollins.

Irvine, Ian Dr., (2010), Jung, Alchemy, and the Technique of Active Imagination, part 3, Mercurius Press, Australia.

Jung, C., and W. Pauli, (1955). *The Interpretation of Nature and the Psyche*. Bollingen Series LI. Princeton: Princeton University Press.

Krippner, Stanley, "The Akashic Field & Psychic Dreams". http://www.ceoniric.cl/english/articles/the_akashic_field_and_psychic_dreams.htm

Krippner, Stanley, "Anyone Who Dreams Partakes in Shamanism". http://stanleykrippner.weebly.com/anyone-who-dreams.html

Lange, Rense, "What Precognitive Dreams are Made of: The Nonlinear Dynamics of Tolerance of Ambiguity, Dream Recall, and Paranormal Belief", DynaPsych. http://goertzel.org/dynapsyc/2000/Precog%20Dreams.htm

Mageo, Jeanette Marie, Toward a Holographic Theory of Dreaming, Dreaming: Journal of the Association for the Study of Dreams. Vol 14(2-3) 151-169, June-September 2004.

Maurer, Leon H., How Unconditioned Consciousness, Infinite Information, Potential Energy, and Time Created Our Universe, Journal of Consciousness Exploration & Research| July 2010 | Vol. 1 | Issue 5 | pp. 610-624, http://scireprints.lu.lv/145/1/85-249-1-PB.pdf

Miller, Iona, 1993-2008, *Dreamhealing*, Asklepia Foundation. http://dreamhealing.iwarp.com/

Misler, Chuck, (2009) New Evidence of a Holographic Universe? http://www.khouse.org/articles/2009/839/

Mitchell, Edgar and Robert Staretz, (2011), "The Quantum Hologram And the Nature of Consciousness", Journal of Cosmology, 2011, Vol. 14.

Nobili, Renato. *(1985), "Schrödinger wave holography in brain cortex." Physical Review A* Vol. 32, No. 6, p.3618-26. Dec.

Pribram, Karl, (1977), *Languages of the Brain*. Monterey, CA: Wadsworth.

Ribeiro S., C. Simões, and M. Nicolelis, Genes, Sleep and Dreams, Chapter 16. http://www.ernestrossi.com/ernestrossi/RIBERIO%20Genes%20Sleep%20&%20Dreams%202008%20Llyod_%20Rossi.pdf

Rossi, Ernest, Art, Beauty & Truth: The Psychosocial Genomics of Consciousness, Dreams, and Brain Growth in Psychotherapy and Mind-Body Healing http://www.ernestrossi.com/ernestrossi/keypapers/ABT%20ART%20BEAUTY.pdf

Rossi, Ernest, (2001-2008), "The Deep Psychobiology of Psychotherapy: Towards a Quantum Psychology of Mindbody" http://www.ernestrossi.com/ernestrossi/keypapers/The%20Deep%20Psychobiology%20of%20Psychotherapy.pdf

Wolf, Fred Alan, (1995), *The Dreaming Universe: A Mind-Expanding Journey Into the Realm Where Psyche and Physics Meet*, Touchstone.

Article

The Value of Dream Work

Iona Miller*

ABSTRACT

Dreams are a form of gnosis, knowledge through direct experience, making them of interest in both consciousness studies and physics. Process work integrates concepts from physics, psychology, anthropology, shamanism, and spirituality into a paradigm and methodology with applications in many fields. Some say consciousness has a cosmic origin, with roots in the pre-consciousness ingrained directly from the Planck time. Process work helps us discover this fundamental awareness and build deeper relationships with our dreams and unconscious. New myths grow in our dreams. As part of the dreamer's psyche, dreams are subjective, but the images in the dreams originate in archetypally informed objective experiences.

Key Words: dreams, REM, dream work, Asklepios, dream healing, shamanism, initiation, psi, visions, holographic brain, energy fields, collective unconscious, nightmares, mythic body, levels of consciousness.

The Value of Dream Work

The archetypes to be discovered and assimilated are precisely those which have inspired the basic images of ritual and mythology. These eternal ones of the dream are not to be confused with the personality modified symbolic figures that appear in nightmares or madness to the tormented individual. Dream is the personalized myth. Myth is the depersonalized dream. -- Joseph Campbell

No one who does not know himself can know others. And in each of us there is another whom we do not know. He speaks to us in dream and tells us how differently he sees us from the way we see ourselves. When, therefore, we find ourselves in a different situation to which there is no solution, he can sometimes kindle a light that radically alters our attitude; the very attitude that led us into the difficult situation. -- C. G. Jung

As we spend a large proportion of our lives in a dream state, a fuller understanding of their implications may prove valuable. Today, there are several prevailing theories concerning the significance and value of dreams. No final statement about dream may be made. There are several approaches to each perspective which is assumed a priori. There are many alternatives to choose from.

Our choice of style in dream work will be determined by the mythemes, memes, or fads we currently embrace. The characteristic attitudes associated with the archetypes will motivate and

*Correspondence: Iona Miller, Independent Researcher. Email: iona_m@yahoo.com Note: This work was completed in 2006 and updated in 2013.

influence our approach to the dream world. Strephon K. Williams (Jungian-Senoi Institute) is one of the foremost proponents of dream work. He outlines a six-point program for continued use:

1. Dialogue with the dream characters, asking questions and recording answers.
2. Re-experience of the dream through imagination, art projects, and creativity.
3. Examination of unresolved aspects of the dream, and contemplation of solutions.
4. Actualization of insights in daily life, where relevant.
5. Meditation on the source of dreams and insight from the Self.
6. Synthesize the essence of dreamlife and its meaning in a journal and apply them in one's life journey.

To offer a variety of other approaches, we will cover theories on dreams and dreaming from Jung's original work, the analytical psychology school, parapsychology, and archetypal or imaginal psychology. Knowledge of the antiquated Freudian system is so wide-spread that no further comment here seems necessary. Jung was the first to depart from Freud's "sexuality-fraught" perception of dreams.

Where Freud saw one complex, Jung saw many. He saw in dreams a gamut of archetypes overseen by the transcendent function, or Self. Analytical psychology amplified and clarified his original material. Most of this work is concerned with the fantasy of the process of individuation. It reflects an ego with a heroic attitude, and proceeds by stages of development. Consciousness, at this stage, is generally monotheistic. It has a tendency to seek the center of meaning, as if there were only One. Parapsychological work done with dreams also seems to reflect this attitude of searching, influencing, and controlling.

In *Re-Visioning Psychology*, post-Jungian James Hillman differs from the traditional analytical viewpoint by stating:

Dreams are important to the Soul--not for the message the ego takes from them, not for the recovered memories or the revelations; what does seem to matter to the soul is the nightly encounter with a plurality of shades in an underworld...the freeing of the soul from its identity with the ego and the waking state...What we learn from dreams is what psychic nature really is--the nature of psychic reality; not I, but we...not monotheistic consciousness looking down from its mountain, but polytheistic consciousness wandering all over the place.

In Jung's model, one major function of dreams is to provide the unconscious with a means of exercising its regulative activity. Conscious attitudes tend to become one-sided. Through their postulated compensatory effect, dreams present different data and varying points of view. Individuation is the psyche's goal; it tries to bring this about through an internal adjustment procedure. There is an admonition in Magick to "balance each thought against its opposite."

Dreams, according to Jung, do this for us automatically. However, there must be a conscious striving toward incorporation of the balancing attitudes presented through dreams (this applies equally to fantasies and visions). Another apparent function for a dream state is to take old information, contained in long-term memory, incorporate it with those experiences, and integrate them with new experiences. This creates new attitudes. Since the dream conjoins current and past experiences to form new attitudes, the dream contains possible information about the future. There is a causal relationship between our attitudes and the events which manifest from our many possible futures.

In studies at Maimonides Dream Labs, Stanley Krippner and Montague Ullman were trying to impress certain information on an individual's dream. They found that an individual, being monitored for dream states, could incorporate a mandala, which was being concentrated on by another subject, into his dream. This led to their famous theory on dream telepathy. Dream symbols appear to allow repressed impulses to be expressed in disguised forms.

Dream symbols are essential message-carriers from the instinctive-archetypal continuum to the rational part of the human mind. Their incorporation enriches consciousness, so that it learns to understand the forgotten language of the pre-conscious mind. The dream language presents symbols from which you can gain value through dream monitoring. You can use these dream symbols directly to facilitate communication with this other aspect of yourself. Should you choose later to re-program yourself out of old habit patterns, you're going to want an accurate conception of what dream symbols really mean.

A symbol always stands for something that is unknown. It contains more than its obvious or immediate meaning. The symbolic function bridges man's inner and outer world. Symbolism represents a continuity of consciousness and preconscious mental activity, in which the preconscious extends beyond the boundaries of the individual. These primitive processes of prelogical thinking continue throughout life and do not indicate a regressive mode of thought. Dream symbols are independent of time, space, and causality. The meaning of unconscious contents varies with the specific internal and external situation of the dreamer.

Some dreams originate in a personal or conscious context. These dreams usually reflect personal conflicts, or fragmentary impressions left over from the day. Some dreams, on the other hand, are rooted in the contents of the collective unconscious. Their appearance is spontaneous and may be due to some conscious experience, which causes specific archetypes to constellate. It is often difficult to distinguish personal contents from collective contents. In dreams, archetypes often appear in contemporary dress, especially as persons vitally connected with us.

In this case, both their personal aspect (or objective level) and their significance as projections or partial aspects of the psyche (subjective level) may be brought into consciousness. A dream is never merely a repetition of preceding events, except in the case of past psychic trauma. There is specific value in the symbols and context the psyche utilizes. It may produce any; why is it sending just this dream and not another? Dreams rich in pictorial detail usually relate to individual problems. Universal contexts are revealed in simple, vivid images with scant detail. No attempt to interpret a single dream, or even the sequence dreams fall in, is fruitful.

In fact, later research by Asklepia Foundation researchers asserts it is more important to journey using dreams as experiential springboards for therapeutic outcomes. In interpreting a group of dreams, we seek to discover the 'center of meaning' which all the dreams express in varied form. When this 'center' is discovered by consciousness and its lesson assimilated, the dreams begin to spring from a new center. Recurring dreams generally indicate an unresolved conflict trying to break into consciousness.

There are three types of significance a dream may carry:

1) It may stem from a definite impression of the immediate past. As a reaction, it supplements or compliments the impressions of the day.

2) Here there is balance between the conscious and unconsciousness components. The dream contents are independent of the conscious situation, and are so different from it they present conflict.

3) When this contrary position of the unconscious is stronger, we have spontaneous dreams with no relation to consciousness. These dreams are archetypal in origin, and consequently are over-powering, strange and often oracular. These dreams are not necessarily most desirable to the student, as they may be extremely dangerous if the dreamer's ego is still too narrow to recognize and assimilate their meaning.

We can never empirically determine the meaning of a dream. We cannot accept a meaning merely because it fits in with what we expected. Dreams can exert a reductive as well as prospective function. In other words, if our conscious attitude is inflated, dreams may compensate negatively, and show us our human frailty and dependence. They also may act positively by providing a 'guiding image' which corrects a self-devaluing attitude, re-establishing balance. The unconscious, by anticipating future conscious achievements, provides a rough plan for progress.

Each life, says Jung, is guided by a private myth. Each individual has a great store of DNA information. It is generally mediated by the archetypes which are deployed by both myth and dream. As you create this individual or private myth, it attracts, if you will, an archetypal pattern and molds itself in a characteristic way (or visa versa). The archetype precipitates compulsive action. It is the motivating factor which may become externalized in the physical world.

Jung notes: "The dreamer's unconscious is communicating with the dreamer alone. And it is selecting symbols which have meaning to the dreamer and no one else. They also involve the collective unconscious whose expression may be social rather than personal."

We may discover hidden meaning in our dreams and fantasies through the following procedure:

1) Determine the present situation of consciousness. What significant events surround the dream?
2) With the lowering of the threshold of consciousness, unconscious contents arise through dream, vision, and fantasy.
3) After perceiving the contents, record them so they are not lost (the Hermetic seal).
4) Investigate, clarify, and elaborate by amplification with personal meanings, and collective meaning, gleaned from similar motifs in myth and fairy tale.
5) Integrate this meaning with your general psychic situation.

Instincts are the best guide; if you are obtaining "value" from your interpretation, it will "feel" correct. Complexes and their attendant archetypes draw attention to themselves but are difficult to pinpoint. We may use conscious amplification of the symbolism presented in dream form. All the elements of the dream may be examined in a limited, controlled, and directed association process, which enlarges and expands the dream material through analogy.

The nucleus of meaning contained in the analogy is identical with that of the dream content. When a dream is falsely interpreted, others follow to correct the error. Preconscious contents are on the verge of being remembered. Just as language skills facilitate new conceptualization, knowledge of the vocabulary of dream symbolism allows closer rapport with the preconscious.

Dreaming is one of the easiest methods of contact with the numinous element, or unknown. To illustrate how archetypes may affect perspective, we will now examine another of the methods for working with dreams and other images. If Freud's view on dreams can be seen as Aphroditic/sexual, and Jung's as heroic/developmental (Yesod and Tiphareth, respectively in QBL), then Hillman's newer "Verbal Technique" might be seen as associated with Hades, Lord of the Underworld or deep subconscious, (DAATH in QBL). This relationship to the image is seeking value, depth, and volume. This method stresses keeping to the image as presented rather than analyzing symbols.

We apply this method to recapture the unknown element because we are thoroughly acquainted with symbols and their. The dream image expresses this if the symbols are not dissected from their "specific context, mood, and scene." An image presents symbols with their particularity and peculiarness intact. Dream presents a variety of images which are all intra-related. Time and sequence are distorted in dream.

Hillman prefers to view dream images with all parts as co-relative and co-temporaneous. This approach to the dream is a sort of metaphorical word-play. The elements of the dream are chanted or interwoven. Repeat the dream while playfully rearranging the sequence of events. Remain alert to analogies which form themselves during this word play. Ruminate on any puns which may occur. As the play unfolds, deeper significance emerges as a resonance.

By allowing the dream to speak for itself, interpretations appear indirectly. This is a method of communicating with the psyche which is in harmony with its inherent structure. In alchemy, it is known as an *iteratio* of the *prima materia.* Its value is evident, according to Hillman. "We do not want to prejudice the phenomenal experience of their unknowness and our unconsciousness by knowing in advance that they are messages, dramas, compensations, prospective indications, transcendent function. We want to get at the image without the defense of symbols."

(1) The archetypal content in an image unfolds during participation with it. *We have found that an archetypal quality emerges through a) precise portrayal of the image (including any confusion or vagueness presented with the image); b) sticking to the image while hearing it metaphorically; c) discovering the necessity within the image (the fact that all the symbols an images presented are required in this context); d) experiencing the unfathomable analogical richness of the image.*

(2) *In this context, 'archetypal' is seen as a function of making.*

The adjective may be applied to any image upon which the operations are performed. This means that no single image is inherently more meaningful than another. Value may be extracted from them all. This coincides with the alchemical conception of the Opus as work. Here the Opus is carried by the dreamwork technique. Archetypal psychology contends that the value of dreams has little application to practical affairs.

In *Re-Visioning Psychology*, Hillman postulates that: *Dream's value and emotion is in relation with soul and how life is lived in relation with soul. When we move the soul insights of the dream into life for problem-solving and people-relating, we rob the dream and impoverish the soul. The more we get out of a dream for human affairs the more we prevent its psychological work, what it is doing and building night after night, interiorly, away from life in a nonhuman world.* The dream is already valuable without having any literalizations or personalistic interpretations tacked on to it.

Hillman ends his "Inquiry Into Image" by stating that the final meaning of a dream cannot be found, no matter how it seems to "click." *Analogizing is like my fantasy of Zen, where the dream is the teacher. Each time you say what the image means, you get your face slapped. The dream becomes a Koan when we approach it by means of analogy. If you can literalize a meaning, "interpret" a dream, you are off the track, lost your Koan. (For the dream is the thing, not what it means.) Then you must be slapped to bring you back to the image. A good dream analysis is one in which one gets more and more slaps, more and more analogies, the dream exposing your entire unconscious, the basic matters of your psychic life.*

This type of analysis seems consistent with the origins of the word. Originally, it had to do with "loosening." This type of dream analysis loosens our soul from its identity with day-to-day life. It reminds us that styles of consciousness other than that of the ego have validity. The soul experiences these styles nightly.

No paper on dreams would be complete without some mention of nightmares. Even though dream is an easy method of contacting the unconscious, it is not always pleasant. Occult literature speaks of a figure called "the Dweller on the Threshold." In Eastern philosophies there are the wrathful deities.

This figure corresponds with Trump XV, The Devil, in Tarot. This seems consistent with Hillman's (1972) attribution of the dream as Hades' realm. The healthy person learns easily to cooperate on his descents into the psyche. The uninformed or neurotic personality is likely to encounter hindrances. These hindrances often take the form of frightening, monstrous, overpowering forces.

Ego-consciousness is not able to comprehend them. When the subconscious is highly activated these images may occur during waking hours and in sleep. This dread and oppression form the basis for nightmares. Pan and his attendant phenomena (such as panic) are archetypal representations of the nightmare.

Pan also corresponds with Trump XV. In the heroic model, as consciousness develops, there is a marked difference in both the content of dream and the dreamer. He gains increased ability to assimilate the charges of energy associated with the dream. The more conscious the experience of the numinous, the less fraught with irrationality and fear the experience is. This holds true in waking and sleeping hours.

John Gowan, in *Trance, Art, and Creativity*, states, "It is this gentling, humanizing process exerted on the preconscious by creative function of the individual which is the only proper preparation for the psychedelic graces." These graces include an immersion of the ego in the expanded context of the subconscious. The ego is then able to return from its experience enriched by the contact.

Contents which might formerly have been considered nightmarish are more fully understood, and the monsters become transformed into butterflies. This attitude toward nightmare is not consistent with Hillman's approach. He does not advocate changing or controlling the psyche. This is, in fact, neither possible nor desirable. He asserts that to enter dream is to enter the underworld, Hades' realm.

Psychic images are metaphorical. All underworld figures are shades or shadow souls. There is no reason for them to conform to the constraints of the ego's dayworld. Soul is the background of dream-work. Underworld is psyche. This relates, therefore, to a metaphorical perception of death. Dreams present us with that different reality, in which pathology and distortion are inherent aspects. We needn't control them, but rather acknowledge their value and depth.

Assuming it is necessary or desirable to control any aspect of dream life, there is a further development of consciousness which enables one to consistently experience what is known as the "lucid dream" or "high dream." (Williams) In a lucid state, there is an overlapping of normal waking consciousness coupled with the dream state. At this stage, one is able to progressively acquire and exercise will in dream states. In the lucid dream, one "witnesses" the fact that one is dreaming, and may take an active role in the unfolding of the dream.

This optional ability is generally associated with the heart-center, or Tiphareth. The heart-center has to do with developing consciousness of the imaginal realm. Rather than control or meddle with dreams, it is more effective to exercise creative expression in waking hours. Many persons pursuing their fantasy of individuation have an outlet through active imagination.

Active imagination is, in itself, an art form. It is generally practiced through a discipline, such as psychology, alchemy, or Magick. It may be dramatic, dialectic, visual, acoustic, or in some form of dancing, painting, drawing, modeling, etc. People who give free rein to fantasy in some form of creative imagination often dream less. All psycho-active drugs also tend to diminish dreaming. In other words, there seems to be a variable ratio between creativity and dream.

Jung made the discovery that "this method often diminished to a considerable degree, the frequency and intensity of dreams, thus reducing the inexplicable pressure exerted by the unconscious." There need be no conscious desire to control or interfere in the actual dream. The ego learns to meet the subconscious on a middle ground, the vale of soul making. The activities and intent of both are harmonized. Staying close to the original image is fundamental.

Chaotic Consciousness & Ego Formation

Enter a space with us. There are fields. Electric. Magnetic. All undefined pure energy-stuff in motion, existing beyond space and time. In fact, time itself is a field--a dispersion of time-stuff, undifferentiated and evenly dispersed. All these fields occupy all space simultaneously. There is another field of stuff -- consciousness -- within these fields.

In certain places in the chaos of intermingling energy-stuff, the consciousness begins to concentrate. As it does, it interacts with other fields and begins to create order (i.e. strange attractors). Fields begin to interact to create matter and time flow. For example, energy plus mass plus electromagnetism is the basis of Einstein's equation $E = MC^2$. The bit of chaos-consciousness that is us begins to form a structure.

Consciousness always strive to take on form. It is still connected to the timeless/spaceless whole, but limits are being imposed on the structure being created. Consciousness is becoming "frozen," concentrated in a limited form. This coming together of fields is the same energy that we call love (cosmological Eros), a primal attracting force.

This represents not only the formation of the human individual, but all other matter in our reality. The interaction of fields, and the formation of a vortex of energy, the attractor, represents the beginning of our consciousness structure. This process culminates in the formation of separate identity, the ego. We can conjecture that in the intermingling energy, somewhere and sometime the beginnings of awareness arise synergistically.

If we trust the dream and consciousness journeys, awareness begins at about this point. It is the first emergence of individual essence from source. In one sense the strange attractors may be the genetic materials, the DNA spirals that come together in an animate condition. It may also have something to do with the inanimate portion which comes together to create the material part of our bodies and beings.

Awareness, perception, and sensing are discrete faculties. Perception is 'seeing through', like the glasses or lenses we put on to see the world. The senses are far more basic than that. We can feel heat in our finger, but the way we perceive that may have many different impacts, based on circumstances and attitudes. Consciousness is base to our awareness.

Dream journeys back to the beginnings of awareness reflect this initial description. Ego arises from this ground state over time through interaction with the environment. Ego is that part of us that is an "I", distinct from a "Not-I". The ego develops a more and more rigid structure over time, as habits and behaviors become "frozen" in the personality.

For healing, parts of the ego need to let go and dissolve the old structure. If you can say there is an "I" or a "Not-I", somehow the ego is involved. This even comes in when you are speaking of the soul. There is soul and there is something that is not this soul. That is really just an ego-model of the soul.

One of the problems with many psychological models of the ego is that they do not take into account what exists beyond the ego very well if at all. Freud's view of the personality includes the ego, the id, and the superego. His view of the superego depicted it as functioning essentially like an authoritarian conscience for us.

The id was considered a mass of unresolved psychic energies (in many cases self-defeating or self-destructive), which nevertheless run us. Freud gleaned some understanding about the id and its effects on the ego through the process of free association. But his conceptual maps were rough. Words are, after all, only symbols. Like a map, they can only provide second-hand information about the complex energy dynamics of the psyche.

We can reclaim from Freud his emphasis on the mythological dynamics of Eros and Thanatos. Eros or love is the essence of the prime attractor, the principle or energy that draws from chaos to create the structure and form which we are. Thanatos is the entropic tendency, the tendency of what is structured to break down into chaotic forms of energy. It works on both our thought processes and physical matter itself. Most people who consider Freud's Thanatos concept see it as negative--perhaps it touches their own mortality complex. In reality, it is not only negative, but probably one of the more important aspects of healing -- this tendency toward "death."

For an ego to change, the old ego must die. Not much attention has been paid in conventional psychological science to that. Most attention is paid to strengthening the ego, to building it up. Jung revisioned Freud's idea of the ego and the imagery of the "Not-I." He came up with the idea of archetypal images from the collective unconscious.

Dreamhealing

In terms of dreamhealing work, one of the most important aspects to come out of Jung's work is his emphasis on the image, and remaining true to the unique presentation of the image in therapy. Jung investigated the transpersonal dimension to understand what existed beyond personality or beyond the self-concept. He stressed the primacy of the archetypal image. The image may be visual or it may be a multi-sensory image. A simple gut-feeling is also an image. It may combine many sensual aspects.

If you can find what the image of the self is, you find that the person's physical and mental make-up takes on the contours of that image. If a person's image of themselves, their very deep primal image is somehow a faulty thing with deformations in it, then the personality will reflect that, and so will the body. It will be disease-ridden and so will the mind, perhaps twisting even the soul.

In the Creative Consciousness Process and dreamhealing work, we have discovered that dreams are shaped by these existential images much like they also shape our lives and destinies. Thus, the surface presentation of the dreams, its symbols and story lines, are doorways that open to a multi-sensory experience of these states. In turn those states of frozen or "differentiated" consciousness can be released or dissolved into the even more primal and base state of "undifferentiated" or Chaotic Consciousness. As chaos theory predicts, a new primal existential image (or attractor) emerges, but one more evolved and more in a state of ease with the flow of natural order. These new, more easeful states provide a new model, a new base for healing the entire organism.

One of the powers of dreams is that they lead us to the image of the self, and that is where the healing generally takes place. Things transform at that level. Observing a series of consciousness states, and mapping these states, we have noticed over time that there appear to be distinct areas of depth that are identified by characteristic imagery experiences.

Dreams can be used in a lot of ways. Examples include using dreams for stepping outside of space and time, to see the future and the past, to visit future lives, to visit futures in this life, because dreams come from outside of time and space. In terms of therapy, dreams can be used as an evolutionary force to take people from a small sense of self and expand them toward a larger image.

To use dreams in a healing sense it is necessary to have an orientation within the dreamscape, to recognize which depth you are dealing with. As interesting as the other uses of dreams are, such as lucid dreaming and interpretation, they do not necessarily lead to healing. They may not even access the state of dis-ease that is troubling the individual, much less be able to re-link essence to source.

In dealing with specific illness there is always a specific image that underlies the ailment. And that is what you look for when guiding a dream journey. And when you find it, you guide the self that this state represents into that state of chaos and dissolution--into a death and subsequent rebirth.

From this chaos, a new image of self emerges. You can go deep in dreams into transpersonal places where there is no sense of self, into true connection with the universal consciousness field. This is the place of chaos, of all and no structure. It is the source of creativity. It is the ultimate source of healing. It is the universal solvent, the panacea. It is the heart of the dream – there and back again.

For orientation within the dreamscape, we use a model which is simply termed an ego map. With it you can find out where you are within the creative consciousness process. It helps you keep your bearings as you guide someone deeper into their journey. It provides clues to help determine if someone is stuck in a fantasy about their belief system or personal mythology. As a map of consciousness this model evolved from our shared consciousness journeys with people as they descend into the depths of their being--their personal underworlds--using dreams as a doorway.

The creative consciousness process flows through the underworld like a powerful river of consciousness, welling up from the subterranean depths. This model is not really specific in terms of psychological theory, but it does help identify the level of imagery you are experiencing. These levels or zones identify levels of ego functioning and development. They are characterized by a quality of imagery and sensory experiences. We can journey imaginally, yet experientially, through the layers of the psyche with this map. This virtual experience affects self-image, emotions, attitudes, thoughts, etc. creating real-life natural consequences.

First we will look at the model from the outside in. Beginning with that which is known, overt behavior, we will move deeper and deeper into the unknown. Moving from the superficial to the transpersonal depths, we notice layer after layer of distortions, and perturbations of psychic energy. These include faulty thought patterns, negative attitudes, and self-limiting belief systems. These are the symptoms of the dis-ease.

This whole process of going deep into the psyche parallels the shaman's journey into the underworkd to find and retrieve the lost soul. It is the natural cure for "loss of self." We seek the lost soul primarily because of the intense degree of wounding in modern culture--alienation. This very wounding has "opened" us to transformation--to healing.

The levels of this consciousness map are not firm-boundaried. They are more like landmarks, familiar zones we have noticed on the journeys. The journey to the underworld, or "the center of the earth," is a metaphor for the depths of the psyche and the wonders we find there. What we find there, experientially, certainly qualifies both as a "treasure hard to attain" and as the retrieval of that which was previously lost or unavailable. It is a process of re-membering.

By repeatedly diving deep and then re-surfacing, we bring into the daylight world very important experiential material focused around the very core of our being. This promotes healing through the imagination. It is a form of meditation, like the alchemical *meditatio*, in which psyche "matters."

The descent and subsequent ascent, going deep into the dream journey and emerging transformed, is a form of death/rebirth, a powerful archetypal theme which is initiatory in character. Before the core, (or soul), is found there are many adventures in the labyrinthine caverns of the deep psyche. There spirit and soul merge in the union of opposites.

Behavioral Zone

How do you judge a person's personality? The first thing you notice is their behavior. You notice how they behave. Whether it is listening to what is said, noticing what is done, personality is revealed.

This is the outer manifestation of self, that being shown and acted out in the social world. It is the level of acting out our scripts and games, the patterns of behavior that reflect the dis-eased primal image. We can use these behavioral impressions to help identify the diseased primal image at its various levels.

Sometimes what a person does and what they say are inconsistent. Right away that tells you about a dis-ease. Generally, you can trust the behaviors more than the words. Watch for incongruent behavior and body language. Body language will tell you the basics of adjustment to life and the situation in terms of openness or closedness, approach or avoidance. From this arise issues of safety, security, acceptance, and therefore self-esteem.

Scripts, games, ego states, and emotional rackets are the foundation of Transactional Analysis. T.A. looks at the behavior patterns and provides a sense of order about them. The basic behavioral patterns are self-reinforcing. People seem to find a way to reinforce their particular pattern whether it is healthy or not. For example, an ego that has come to believe, think, and feel that it will be rejected by others will engage in behaviors which make that a self-fulfilling prophecy. That helps maintain the (distorted) ego structure.

The client's standard games are bound to come up in therapeutic interaction. That is how they relate, cope, and assimilate their experience. You know they are bound to wind up playing the same games in therapy that they play everywhere else. They are going to try to prove the same thing. This is what Freud referred to as the transference, a projection of either/both positive or negative parent.

When the energy flows toward the client, it is called countertransference. It is an unconscious, automatic process. For example, clients might come saying, "*Well, you can't prove to me that this therapy works.*" If you respond, you have fallen into their game. As soon as this happens, both become involved in a power struggle.

Clients who try this game essentially use it in most of their lives. This skeptical, confrontive, challenging attitude creates problems in all areas of relationship, but can generally be managed in short business exchanges. In this example, just asking the question is the expressed behavior. And it will reflect, however abstractly, the basic dis-eased primal image. Within the symptom is the shape of the disease. In the dream, this level often shows up in the interpretive level in dream work.

In Gestalt experiential dream work, in the experience of being the parts of the dreams, one often notes how well the dream parts and their interactions reflect the dreamer's outer maneuvers. Behavioral psychology tries to deal with change at that level, as do many other superficial psychotherapies.

In dreamhealing and creative consciousness work, however, it is merely noticed as providing information about the shape of the diseased consciousness state that we will eventually encounter in the depths of the psyche. It acts as a map to identify the shape of the personality, to help us guide the dreamer beyond this level.

Transactional Analysis provides an excellent means of conceptually understanding this level of the personality. It provides structure to help identify the repeating patterns that are the behavioral level, reflecting the dis-ease within. These are probably just symptoms of the dis-ease you will encounter.

The literature of T.A. covers behavior patterns quite well. Perhaps the best single source is BORN TO WIN by Muriel James and Dorothy Jongeward. Conditioned behaviors are cataloged in behavioral psychology which concerns itself with basic instincts like fight/flight, how habits get formed and broken, and manipulation of behavior through punishment/reward systems.

Thinking-feeling Zone

How we behave is largely shaped by how we think and feel about what is happening in the environment. This means "feel" in the emotional sense as well as in the opinion sense. It is a deeper zone than the behavior, which is mostly acted outside the self. This zone lies beneath the surface, a zone of thoughts and emotions, competing and dancing with each other to create patterns that shape our behaviors, but are themselves merely reflecting the even deeper patterns of dis-ease. In this aspect the dream journey reflects the experience of delving deep into the structure of a fractal image, where each layer repeats the basic structure of the levels above it and below it.

The probing of the structure of dreams and personality is probing the nature of chaos itself. Cognitive-emotive therapies, such as T.A. and Gestalt, most of the psychoanalytic and Humanistic therapies, and techniques such as affirmations work within this level. They often try to produce changes at this level.

Models to explain the structure and function of this zone proliferate, and often contradict one another. But in the creative consciousness process, it just provides the dream guide with another clue, another perspective on the shape of the deeper dis-eased primal image. It adds another sensory configuration to the patterns and shape that will eventually identify the primal dis-eased state to be encountered further on in the journey.

It is rate that the seeds of disease originate at this level. This level is encountered and revealed in dream again at the surface presentation level, and slightly below that. It is revealed in the emotional content, plot lines and first levels of symbol experience, as for example, encountered in Gestalt work. The imagery is discreet, that is this dance of emotions and thoughts is experience with easily recognized patterns of imagery--cognitive and emotional process.

The first step in a dream journey is to re-experience the dream. In that re-experiencing, these emotional, cognitive and action sequences are experienced by both the guide and the dreamer. These sequences usually reflect the behavioral and script game patterns noticed prior to the journey, and round out as well as reinforce the emerging sensory image the dream guide is developing out of the state of disease.

Dream interpretation deals with this level of the dream and is a well-developed art. Surface symbol manipulation, as practiced in techniques such as Gestalt, or dream psychodrama also explore this level and promote change at this level of personality. But in the creative consciousness process, the dream guide only notices this level of experience and these dynamics.

The patterns experienced at this level are models that will help identify deeper patterns and eventually the "source." Let us present an example, an illustration of a typical journey: In the dreamer's dream, he is continually frustrated as helplessly he is drawn into an impending disaster that it seems cannot be averted. At this level of presentation, the dream is reflective of the dreamer's outer emotional-thinking dynamics and patterns.

In outer life he feels like a helpless loser who is constantly beset by crisis and disaster. He is constantly in no-win situations. Most dream therapies, analytical, interpretive, etc., would attempt to deal with changing these patterns and dynamics at this level by whatever therapeutic techniques they espouse.

Even experiential Gestalt would seldom venture any deeper than exploring this level of dream experientially. But in the creative consciousness process, the dream guide instead co-experiences these patterns and energies. If the mind is involved at all, we may speculate that the deeper experiences and patterns of disease may exhibit a sense of "stuckness" or something like that. Having noticed, and speculated, the guide then lets go to journey even deeper into the dream and the psyche.

Belief Systems Zone

The next zone we encounter in our journey into the ego, and what is the same thing, into a dream, is the belief system zone. The answer to "What shapes the thinking-feeling patterns?" is what we believe about self and the world. This is still a somewhat intellectual zone in that most beliefs can be succinctly stated with a few words or sentences. But beliefs arise also from much deeper levels of sensations. Deep feelings, senses of credulity, rightness, wrongness, all help identify the boundaries and shapes of our beliefs.

Of course, in the dynamics of life, experience of this zone is a dance with the emotional-cognitive level. That is how it affects the pattern of the dance. But functionally it can be justified as a discrete zone. Again, many conventional therapies from T.A. to Freud deal at this level of ego functioning. Indeed, changes in belief systems are an integral part of depth therapy healing. In T.A. this is the level at which deeper script and existential patterns are revealed and re-decided. But seldom does the diseased state originate at this level. However, most therapies attempt to deal with the disease at this still somewhat symptomatic level.

The creative consciousness-dream guide again just notices these sensory patterns. They provide yet another, deeper and new way of identifying the essence of the primal dis-eased image that will eventually be encountered. At this level the imagery as we approach the boundaries of the beliefs, or the area of the known often takes the shape of dangerous paths, or other fear-filled images. At this deeper sensory level, images appear such as walls felt as solid barriers, or ugly sensations and colors, or perhaps odors or sounds, monsters or evil creatures of cold and dark. These are energies that keep us in bounds, and trapped in our belief systems. At this level the dream journey is indeed the hero's journey.

The dream guide must lead the dreamer through fears to even deeper levels. It is often sensed as a journey into death or worse. For example, the frustration in the dreamer's dream might suggest a deep red color, which when experienced takes on the feeling of sticky pools of blood on a cold tiled floor. In essence, the message and belief is that it is death in a cold sterile place to go beyond this point. In essence the outer message is that it is better to be a failure and be stuck than to die.

On this outer level the dream guide might speculate about a child having these deep feelings and deciding in essence to never surpass their father because...well, it doesn't really matter why. The point is that it is an experiential memory of a reaction to something that happened once. What that is does not matter to the dream guide. What is important is that this represents the energy or psychic boundary that keeps this person trapped in their belief system of helplessness and failure.

We are getting closer to the primal image now--to the state of disease. There is an essence of woundedness and death in it, maybe even a sense of drying up and becoming sticky. But again, to the creative consciousness-dream guide, it is only speculation what this all means. In fact that is often far from mind, as the quest is to push on ever-deeper to the source of this pattern-form. If anything, it is now stored in the guide's deeper senses to emerge later to support the intuitive feeling of "fit" that identifies the deepest level of the diseased primal image, when it is reached.

Personal Mythology Zone

The next zone can be best characterized as the "Personal Mythology" or Archetypes Zone. This is the level from which our belief systems are formed. To the ego, there is still some slight component of rational process. For example, this level is often revealed by the fairy tales (personal mythology) that underlie the scripts (belief patterns) in T.A.

The imagery in this zone is, in most cases, significantly more superficial than the archetypal images suggested in Jungian psychology. In a sense they are distorted versions of the archetypes (complexes). These are the distortions caused by the organism's very early interactions and experiences with its environment. In some cases they are direct representations of archetypal energy patterns (remember the strange attractors we discussed earlier).

We are very close now to what forms us out of chaos. The archetypal patterns or myths adopted are the ones which most closely reflect the organism's shaping experiences. Stanley Krippner and David Feinstein have described this level of consciousness and its impact and role in the human condition in their book, *Personal Mythology*. It does not conflict with what we find in dreamhealing experiences in any important way. They speak of five stages in the process of integration with one's evolving mythology.

Their work is one of the first serious forays of traditional psychotherapy into the mystical realms, other than anecdotal reports. They are actively working at the mythic level with therapeutic interventions. The five stages include the following: 1). Identifying areas of conflict in the person's underlying mythology. 2). Bringing the roots of mythic conflict into focus. 3). Conceiving a unifying mythic vision. 4). From vision to commitment. 5). Weaving a renewed mythology into daily life. Krippner and Feinstein have used Graywolf's life story to illustrate their model on many occasions, including Krippner's *Dreamtime and Dreamwork* (1991).

It is probably no coincidence that this closely resembles the four stations of the Medicine Wheel: identifying the problem, letting go, new vision, and empowerment or actualization. This is the archetypal healing model operating at the mythic level, no matter how it is stated. We don't seem to automatically jump into a belief system from an experience. It has to go through a mythologizing process, and thereby become entrenched as an image. That is an important part of how the image gets in there. We turn an experience into a belief through the process of mythologizing. The mythological layers form a sort of border between the ego and the personality, separating it from the deeper collective psyche.

The mythological layer is the boundary layer between the personal and transpersonal. That is why the archetypes are so clear there, yet they are also in a logical context, story, or drama. It is the melding of rational and irrational. Mythology is a precursor of evolution. New guiding myths are arising today, such as the new myth that "personal power arises within." We follow an image or myth of who or what we expect ourselves to become. This is our existential myth. Identity crisis follows if it doesn't work, creating an evolutionary crisis.

This level of consciousness is profound and indeed at this level the guide may encounter the primal diseased state itself. Most often one must forage deeper, but sometimes it rests in this level. These consciousness states/images are frozen into form very early in life, within the first year or so. They manifest as deep sensation and sensation patterns and are a level of memory experienced pre-verbally, or at barely verbal levels of cognition and experience.

In a way they can be viewed as the time when the organism is beginning to shape its "self." The organism has rudimentary experience at the sensory level, but no cognitive-emotive existence. It seeks the archetypal energy forms rising from its chaotic origin, and selects the ones which most accurately match its experience so far. It modifies those archetypes to match its experience and this provides a strange attractor that becomes the nucleus of personality.

This zone holds very little "mind" content and is purely sensual in nature. The visual imagery (if there at all) is simple, perhaps surrealistic -- disembodied eyes or faces, frozen statues, pools of molten red lava, animal faces, jaws full of sharp teeth, etc.

Archetypal Zone

But still deeper strata exist--a deeper zone of psychic energies and patterns that represent memories of even more primal levels of consciousness and experience. When we reconnect with them experientially, we re-member our deepest self and heal our fragmentation. Here we find experiences of the Womb, back even to the dance of energy, matter, and consciousness at Conception. These images are close to the stuff of our creation, the primal chaos or consciousness field that seems to underlie all matter.

This zone is one of archetypal energy waves and patterns, existing on the edges of infinite creation. In this zone the imagery is beyond surreal. It is psychedelic or mystically sensual, much as described by individuals in the deepest of L.S.D. experiences, or during moments of ecstatic healing, or religious experiences. These strata are cataloged extensively by Stan Grof in his works on LSD therapy and Holotropic Healing. Here are revealed shifting, dancing energy patterns that sometimes only suggest forms, or may assault the senses -- deep whorls that suck one down in spinning spirals, black holes in black space, gray clouds of nothingness. Senses are extraordinary and seemingly infinite in variation (including the controversial psi phenomena such as extrasensory perception (ESP) and the sensory melding of synesthesia).

If this zone is clear, the dis-ease exists at more superficial levels. It is a zone of great ease, and an experience of timeless flowing creation. It resonates as deep rightness and peace, ageless antiquity. If the dis-ease exists at this deep level, the experience is similar to the above, but somewhat modified.

For example, a deep red stab wound with a black center might become a swirling vortex pulling the dreamer and guide into a blackness. It is cold and empty, and the spinning of the vortex has dismembered the dreamer. In fact, he has experienced a sense of being disintegrated. In this state of nothingness a throbbing sensation of pulsating red becomes a sphere of softness surrounded by a shell of resistance. At the same time spikes are puncturing the shell. Becoming both the punctured and the puncturer leads to a deep-felt sense of flowing togetherness and peace. There is acceptance of conception and creation as deep-felt senses adjust to this yielding.

Chaotic (or Creative) Consciousness

The zone of chaotic consciousness underlies all of the foregoing. Within our theory this is the level at which all structure dissolves and from which all structure comes. It is the "universal solvent" of alchemy, the liquid form of the Philosopher's Stone. It could, in fact, has been called by many other names. It is a sea of universal consciousness, timeless and infinite. It is chaotic consciousness, a level of being and energy that is virtually random and unformed. It is a state of pure creative energy with infinite possibilities. It is the Tao. It is the timeless, spaceless quantum leap. It is a higher order dimension of self-existing in hyperdimensions or virtual realities. It is the healing, creative HEART OF THE DREAM. It is the selfless Self.

What is the imagery? To borrow from Taoism, "*the Tao which can be described is not the Tao.*" It is remembered on returning as being all that has ever been, all that is now, and all that is yet to be. Here lie buried memories of conception, the instant we begin to develop a "consciousness" of the self from the collective consciousness. In so many of the conceptions we hear about in dream healing, the energy is wrong. For instance the mother is being raped, or nobody cares, or it is an accident, or father was drunk, etc.

Conventional psychology has not dealt with this phase as an experience of trauma very much. Even the alternative therapies focus more on the birthing experience. Eric Berne, founder of T.A. used to pose the question, "*Make up a story about your conception.*" That was part of the script questionnaire.

You can discover a lot from that simple exercise. We are only beginning to realize just how much of a person's form and structure, not only genetically, but also psychologically, comes out of that initial experience. The dreamhealing method has a big advantage in that it views memory differently than most people view it. It is more than the conscious process of recall. "*I remember this happened,*" is usually a visual or auditory memory. When you begin to leave that perspective, you can perceive that the deep memories come to you in other forms. Genetic memories come to you in different forms. They are not discrete memories. They are sensory imagery, and structural characteristics. Then memory expands to include a lot of recorded experience you just cannot get at in other ways. Just to process, re-experience, and reorder those memories is therapeutic.

It means going into primal chaos to begin the process of reformation from the most fundamental beginning. The process of conception parallels chaos theory in that these initial conditions are very critical, and slight changes in those conditions can bring on the exponential disruptions of the "butterfly effect." Chaos theory uses the metaphor of a butterfly in the orient flapping its wings thereby causing a gigantic storm on a far-away part of the globe. During conception, we have the initial chaotic conditions which begin forming the initial structure. The slightest things that are wrong here may have horrendous effects down the line.

Having descended to the formative depths, we can now begin to ascend through the consciousness map through the typical stages of development.

LEVEL 0: THE SPACE/TIME CONTINUUM AT CONCEPTION

Conception occurs with the interaction of the space/time continuum with collective or universal consciousness. The ephemeral soul enters real-time experience as a tangible entity. Yet we are still totally immersed in the unified web of awareness. Direct experience of this level is a true sense of oneness with all. It is not just a metaphysical idea, but a real field that exists -- a permeation of space/time with consciousness.

The collective or universal consciousness may be seen as an all-pervasive consciousness that exists through all of space and time. Each one of us is a part of that, and connects intimately with that. Jean Houston has called this the I AM experience. Our consciousness is only a manifestation of this larger consciousness. It is spoken of as a union of opposites, or God, Unity or the Tao, etc. It exists in stars, in ourselves, in all things. It is also totally undifferentiated. In it there is no sense of separation of self, or anything else.

LEVEL 1: COLLECTIVE OR UNIVERSAL CONSCIOUSNESS AWARENESS

This level where we are ALL ONE is a very healing state of consciousness. Consider the idea of the space/time continuum, with the three spatial dimensions plus the fourth dimension of time. If our consciousness is trapped in space and time, we essentially live a Newtonian cause/effect life. This happens which inevitably results in that happening. At surface levels we experience a causal world. At deeper levels of the psyche we can make the quantum leap in consciousness to a seemingly timeless/spaceless realm where we can experience ourselves differently. In this acausal awareness we are reconnecting our essence with our Source.

Consciousness exists like an ocean. Jung spoke of it. The mystics call it the Father (or Mother), the Source, Great Spirit, etc. Communing with this energy, experiencing this state of consciousness, was the practice of shamans from the beginning of human history. They developed many techniques for "getting there."

Consciousness always strives to take on form, and spirit urges us to cast off gross form and return to primordial unity. What creates the space/time continuum may be the interaction of consciousness with the other fields that exist, such as the time field, EM fields, gravitational fields, and the "strong" and "weak" force within our atoms. These fields are virtually inseparable; they nest within one another. At the level of the still-elusive unified field theory they are one -- and we are that. In fact, Jean Houston terms this experiential level, WE ARE.

No matter how we define it, this is the core or source from which we come. What happens is that as we enter four-dimensional space, we develop increasingly complex dynamic form and structure over time. These energies crystallize around the nucleus of consciousness interacting with the space/time continuum, and perhaps other (hyperdimensional) dimensions. If you can consider a dimension beyond that of space/time, it is without form -- a vast non-linear, pre-geometric ocean of disintegrated virtual energy -- pure potential. It is chaotic – a chaotic dimension – a chaotic consciousness.

Everything is de-structured here, disassembled. It is hard to envision form or structure existing beyond that. From this infinitely vast ocean of potential arises a wave. It differentiates like a wave on the ocean -- a "standing wave" in four-space. As this consciousness differentiates and begins to enter local reality, we can call it the soul, if you like. We are not solid matter at all, as the materialistic view teaches us.

Rather, we are dynamic wave fronts in the ocean of the continuum. At the moment of conception, the organism begins to exist in the space/time continuum, in the physical world. It begins to be trapped by the deterministic laws of cause and effect, and is still subject to the bizarre-yet-deterministic laws of chaos. Within these parameters, the entity begins to have experiences which develop awareness. The organism somehow stores this "pre-experience."

LEVEL 2: ENTRY INTO SPACE AND TIME

With entry into space and time, the unconsciousness wave enters the realm of material reality, and hence duality. The genders merge in act of love (or merely sex). The dynamic interaction creates an attraction (an attractor) or agitation in the virtual energy of the collective consciousness, which becomes a "wave" in the ocean of consciousness. Conception takes place as the wave finally enters into material reality--sperm meets egg.

The wave might be considered an individual soul, but the embryonic soul has the dual nature of being purely collective at this point yet invested with the potential for individuality through ego and personality. The divine collective interacts with material reality and begins the process of forming the physical and psychic self. Since at this point, we don't have much of a body, or mind, or form, or structure, this early experience cannot really take on much of a sensory memory. We can't remember much of how it felt, or how it sounded. But we can remember it in terms of energy itself.

So memories of the very early images and impressions are stored in terms of energy. With this conception we envision a metaphorical rebirth, a rising from the depths of the underworld. We have found the lost soul – the lost self. This process of remembering the deep self, the core self, heals dismemberment. By re-membering, we re-create, and re-new ourselves.

Conception takes place when the attraction of love draws a wave of consciousness into interaction with space/time, as the sperm symbolically pierces the egg. "Love" in this sense is the power of primal attraction, mythologically the cosmic Eros, not necessarily love of the partners in the sex act. Love is a creator, healer, and unifying force in all human experience, spanning both scientific and mystical reality. Love energy is "sex-red," similar to the red of Buddhist reincarnation theory.

LEVEL 3: PRE-SENSORY, GENETIC CONSCIOUSNESS

Conception is the awakening of genetic consciousness. Its dynamic mandate is set in motion at that point. At this instant of creation, prenatal awareness begins and memories begin at the molecular or cellular level. The collective consciousness is given the material with which to create its body, but it is now restricted by the laws of material reality.

At the same time, energy experiences (the situation in the womb) begin to form the psychic body or the ego. At the moment of conception, the collective consciousness begins to exert its influence and create a body and an energy, and/or psychic shield to give it both protection and to allow for the perception of and interaction with material reality.

Thus, two "bodies," physical and psychic are now forming in the womb. There are pre-sensory images. These appear at the deepest levels of dreamwork as a totally expanded sense of self -- no boundaries or limits. Another way to say it is that consciousness first intrudes and then limits part of itself to the constraints of space/time. Input to the fetus is very basic at this point. Nutrient input yields a sense of getting or not getting. Emotional input from basic chemicals borrowed from mother and mother's body sensations are stored as pre-sensory images.

If mother's chemistry is toxic it sends the message that the physical world may not be a safe place to be. If mother is an alcoholic, the fetus is exposed to poisonous blood. It damages self, causing it to see the world as poisonous. It is a sensory image based on the whole environment being poisonous. Or, if the moment of conception is a moment of hate, violence, or rape there is going to be a lot of energy attached to that moment of conception. The very creation of self, the very act of formation of self is based on that, whether or not these circumstances of conception are later made conscious or not.

The process of personality formation covering soul is like an oyster forming a pearl around an irritant. The subsequent layers of personality that overlay on top of that take on these very early shapes. Early hormonal reactions of both mother and child are also experienced at this level. If you guide a person to this level so they are experiencing it directly, they can actually affect their hormonal balance. If you start with dreams, they sometimes heal physical problems, too. When their awareness enters this level, you can help them to change some of these images. They reenter this state of consciousness and come back out to build a new sense of self, a new personality, even a new body and chemistry.

If you penetrate to the primal level, you touch back into the collective matrix. In the womb there are experiences which leave imprints, such as chemical memories. As the fetus grows, it develops early sensory apparatus -- nerves -- so that these memories are stored now as basic sensations. These are pre-visual sensations. This is usually a very basic visceral or gut-level perception. It signals the awareness of comfort/discomfort. This awareness provides the earliest sense of value for physical existence. Is it nurturing and friendly, or does it abandon and reject the well-being of the fetus? Is initial sensory experience comfortable or uncomfortable? During the time in the womb, sensory abilities become more and more complex.

So, the pre-sensory level takes on the form of colors and heat, without any visual form or imagery. The impression is of drifting things, maybe a void. Very subtle sensations, color, and elusive impressions characterize this level. Some clients have trouble describing what they are experiencing in the dreamwork when they reach this deep level. It is based on very early experience. It is a conceptual experience, a womb-like experience. It is so basic it comes before the brain is even formed, so it is totally raw. The genetic consciousness is acting on the genetic material supplied by the mother and father. Following the laws of genetics, it creates a physical being which can sense and interact with material reality. As the embryo grows new senses are added to visceral awareness including sound, sense of color, and images, etc.

Memories of space/time consciousness are stored in this way. It is still an undifferentiated sense of being. The psychic consciousness is creating the various aspects of the energy body. They are connected to the physical body through such structures as the ego, the astral body, and the meridian system. Our conscious awareness is usually limited to just a segment of the ego, but the ego reacts with material reality also.

Around the second or third month of life, we develop nervous systems which exhibit some discrimination. Then the pre-sensory images evolve toward sensory images. A new way of storing experience is developed. In process work, people can explain their experience more in terms of the known senses, excluding vision. These sensory images still have a lot to do with throbbing, pulsations or sounds, and colors like pure red and black. This is very often the fetal heart or placenta beating rhythm. Experiences of this phase can impact how a person forms or builds a personality in a very profound way.

LEVEL 4: BIRTH AND SENSORY IMAGES

Birth occurs during this phase and the sense images increase in complexity. The undifferentiated slowly becomes differentiated into images of self-world as shapes, colors, tastes, motions, feelings, emotions, sounds, and acts. This is how memories are stored. Birth imagery includes tunnels and caves, feeling pushed or expelled. As the differentiation sense grows, the perception of I and Not-I, and sense drives come into awareness. Comfort/discomfort perception continues and becomes even more acute. Another impression becomes perceivable which we will term empathy.

Images begin to take definite shape via sound, taste, odor, etc. The birth experience may be stored deep in the subconscious, and sometimes becomes the subject of a dream. Frequently these dreams come up in midlife when psychological re-birth becomes an inner urge or drive of personal evolution. The original birth experience can characterize or color the rebirth experience, whether it comes through dreams or process work. For example, one client who was delivered by Caesarian section had the following dream, which for her amalgamated the images of both birth and rebirth.

"CRAWL-AWAY": I'm with a group outside looking at a house. We watch a person struggling to get through a hole or opening in the foundation. There are lots of comments about WHY he's having such a hard time. We go inside and look around (apparently there's some problem). For some reason the men in the group are going somewhere, in or out of the house, to do something. Something happens (explosion or earthquake or something) and the problem is much worse, and there is little or no light. I tell them that I will go and look for the/a way out, the problem or something. I go down a hallway (with another female, closely, quietly and apprehensively behind me)... for some safety reasons or something. *I/we have some kind of illumination (not very bright). The hall changes into a smaller passageway and then very small, like the crawl-way beneath a house, and it gets smaller all the time. The one behind me gets more frightened and pushes closer making it a lot harder for me to move along at all. I come to a slight turn on my right and find that the regular way out is blocked by cement blocks and rubble. Passage through there is impossible, and there is absolutely no way to turn around and go back!!! The one behind me is so close and won't move back at all. We remember that WE, the group, have something to do with blocking it for some safety reasons or something. The passage is so very small at this point. I noticed that there is a small crack in the foundation to my left and behind my shoulder, but I've passed it a little and it's sooo small I'm not sure that my head will even fit through it! For the first time I'm scared! The one behind me crowds even closer if that's possible, and makes it even more difficult! WE CAN'T GO BACK...OR FORWARD!!! There is no more illumination...our only chance is to go through the crack. I squirm around and maneuver so that I can try to squeeze out. I manage to get my head near the crack and put it up to it...and all of a sudden I'm in the bright light on the outside. I look back at the crack and remember the other having a hard time getting out. The first thing that came to my mind and feelings was that I had just been born, again? I was in an adult body all the time even when I got back into the light, on the outside.*

LEVEL 5: THE DEVELOPING SENSES

During the first months of life, the eyes and ears develop completely, and the brain discriminates more and more. As time goes on, we get more and more from the senses, learning how to use them more in the outer world. We then begin to form sensory imagery. These images begin to put things together, to create associations and natural metaphors. The organism strives to define self. Primal images of world-self begin to form.

Body ego and much of personality ego will reflect this shape, (perhaps contour is a better image). The edges may be softened by later experiences but it still will determine the base shape of the personality/body. This level is the basis of physical disease and susceptibility. It is the level of predisposition. At this time using a combination of genes, even basic body chemistry, processes are set in motion which may result later in cancer, heart disease, etc. Here it becomes established and becomes a strong potential.

At this stage of development, material reality is filtered through both body and ego-mind. At first, it is mostly just impressions of images of the immediate environment. It forms a layer of sensory images of the world, a base range, against which all future ones are checked.

Clients often describe imagery of this level as a paradoxical union of opposites, such as a full/empty or hot/cold feeling. There are also many extrasensory impressions from this level, which are often contradictory. Language is virtually inadequate to describe these dichotomies. More structuring of the original chaotic consciousness of pre-conception orders a personal reality as the baby develops. Senses become discrete perceptions. Sight, taste, and odor become differentiated rather than melded in synesthesia. There is finally a glimmering of the separation of mother and self.

This takes place at the pre-intellectual, pre-conceptual stage, but the awareness comes into being and is the seed of ego development. Some people's entire basis for a fear of the world is the image of an angry face. When they go back, they eventually discover it is usually a parent's angry face, maybe just impatiently telling the child to go back to sleep. But the child sees that annoyance, senses that, and internalizes that anger. A small child stores that impression, as a physical and psychological (psychophysical) gestalt. It becomes encapsulated in that vision of the angry face, which seems to reappear in later situations, creating the same automatic response of disproportionate fear.

All angry authority figures somehow become that angry parent. Each repetition reinforces the image. Later, the person walks in to see the boss, and he's got an angry face, so immediately the individual folds. Psychology is good at sometimes reaching down to this level to resolve those issues by helping the person realize there are alternatives.

There is usually an image that is stored in the mind that is a complete image that has emotions attached to it. These are olfactory, visual, auditory, and kinesthetic sensations that encode its essence. It involves more aspects of your total sensory being, or sense of being. These images can be processed with NLP techniques such as the "re-frame" and "change history," but the results are limited and sometimes do not "stick." Changing imagery at this level is not necessarily the whole answer, because it is just a reinforcement of more primary belief systems.

LEVEL 6: MYTHOLOGICAL LAYER

Underlying the ego layers of personality is the mythological phase of development. It directly underlies the personal belief system, and is instrumental in its formation. The other component

is experiential--the interaction of the personal and transpersonal. Much of the appeal in myth derives from the fears and fantasies every child experiences as part of the way he defines himself. This is also the level of fairy tales and heroic epics. Our role models and cultural heroes glean their appeal from their identity with the mythic characters.

The structure of the heroic, upwardly-striving ego also resonates with this imagery which is influential in its formation. One of the realizations we need as modern people is that this heroic, perfectionistic, overachieving ego model may rob us of our humanness. There are many other, gentler archetypal tales. A favorite fairy tale can condition an entire life. Many a Cinderella laments that, "*one day my prince will come*." This is the old rescue fantasy.

Our youth-oriented society asserts it will "never grow up," and rejects wholeness by disowning its shadow like Peter Pan. Another example of the mythological layer is the tale of the "Emperor's New Clothes". Embarrassment might be encoded something like this: "*Somebody pulled a fast one on me. Here I am walking about naked, and someone pulled a fast one on me.*" And that is how we store it; as a child that is how we experience that. This is why fairy tales enchant children through identification with the metaphors behind the story.

The identification begins with personal experience and is validated in the story -- *"Hey, that's what I felt, thought, imagined, believed."* For example, this embarrassment ("bare-ass" ment) might have its precedence in earlier childhood when parents insisted a child perform. They may want something simple, like saying DaDa, but the child can become the object of derision.

When he can't perform, and receives ridicule instead of praise, the small child may feel betrayed, exposed, and abandoned emotionally. Many incidents repeat the essence of the experience, basically confirming the more basic existential beliefs. How many adults today would freeze with fear asked to speak before a small group of people, because of shaming in school? These levels tend to be stages of memory stored in images of the senses we pay the most attention to.

As we get deeper and deeper in the mythic script, we begin to get into other senses than the normal five we use to deal with the outer world. This is the psychic aspect of psyche, and involves phenomena like telepathy, clairvoyance, and synchronicity. In dreamhealing practice, these are spontaneous aspects of the co-consciousness journey. They arise within a no-boundaries or no limiting expectations condition.

LEVEL 7: BELIEF SYSTEMS AND INTELLECT

In the more surface level of belief systems, beliefs are stored in the form of actual memories, stories which are almost mythologies. They become mythologized over time much like our real culture heroes become the stuff of legends. To continue our previous example, the memory of being laughed at in class can develop into a memory of the world as a place that is always going to ask me to do things and then laughs at me for doing them. This embarrassment can lead to introversion or avoidant behavior, and negative self-talk about self-esteem. Images stored around that memory become a whole belief system about "who I am" and "what the world is," and "how I behave."

At the sensory level colors, sounds of throbbing, warmth/cold, comfort/discomfort are typical experiences we hear about in session. As we continue to grow we add intellect to this imagery and begin to form belief systems. They are our minds' way of making sense and putting things into a structure. A desire for order is basic to our survival instinct. Structure gives us an easier way of dealing with things.

Belief systems form as we begin to make a structure with basic existential beliefs and later fairy tale beliefs. As abstracting ability begins (programmed genetically), the images take on aspects of a dynamic story with interactions. These are fairy tales and basic personal mythology--archetypal shapes and sequences. This is the level of Freud's id. Identity is a key issue here. The sense of who one is leads directly to emotions, thought patterns, and behaviors. Of course, behaviors always feedback and reinforce the beliefs, which reinforce the behaviors, ad infinitum unless there is intervention.

As perception sharpens and words and ideas are processed by the brain to add to sensory impressions, word images are formulated and create the very first and most basic belief system against which all future experiences are compared. In later life the touchstone of familiarity is generally chosen over well-being, so this imprint is extremely important.

In T.A. this layer of beliefs is known as life scripts. This thinking activity later becomes a resource for the adult self. Conceptualization and generalization begins and images of experience form the foundation of belief systems. As the brain begins to develop abstract ability, it tries to organize experiences. First comes the level of personal image, the mythology of infinite self-god, a solid world relationship.

As the intellect develops still further ability to abstract, there is emergence of a belief structure about self. This is the earliest form of Script decision. This level summarizes all experiences of self-world interactions. It may take on attributes eventually of several "intellectual" belief systems as intellect cannot describe the entire sensory gestalt by a single belief.

Belief systems give rise to how we react (feel, think, and emote). As we perceive what is around us we compare it to our stored impression of what reality is, what is I and Not-I. We determine its nature, make a judgment, and this determines how we think-feel, and this in turn determines how we behave. These games and patterns manifest on the ego level.

LEVEL 8: EMOTIONS

The unique emotional reactions of the individual are directly based in the belief system. It gives rise to certain emotional patterns which are coupled with or complexed around each belief. For example, a "mother-complex" conditions all other relationships and keeps the inner child infantile. A "father-complex" may inspire a rebellious attitude which also creates dysfunction in other areas of life. Each belief generates an emotional response that surrounds it. This forms the core layer of the ego.

We can speculate that all experiences that separate us from universal self are uncomfortable, chaotic, painful, and fearsome to some extent. Sorrow, pain, and suffering are inherent in the nature of a self-reflective consciousness. Both psychologists and mystics share this notion. This pain of alienation leads us to question, wonder, and experience awe. Fear "freezes" us rather than allowing our energy to flow in a balanced manner. In fear, the I is hurt by the Not-I, even at the earliest ontological point. Pain helps us define I and Not-I; a hot stove lets us know right away.

Circumstantial pain may not be useful at the time. But pain can lead to fear, which leads to a belief which complexes as a fear of pain. In dreamhealing the remedy is to go through the fear and pain to get to the heart of the multi-sensory image. Past the fear and chaos is a peaceful, calm center, a special place, a transcendent state which is naturally healing. In Transactional Analysis this layer is represented in the racket system and emotional aspects of games.

LEVEL 9: THOUGHT PATTERNS

Almost back to superficial reality, we find that emotions in turn give rise to the thought patterns that cluster around these emotions -- belief clusters (complexes). This is the next ego layer of thinking, which may not be an entirely separate layer from emotions. They interact in lock-step. For example: Through the senses I trigger 'belief system A' which triggers the set emotional response. At the same time, by my intellect, I give myself the thoughts that rationalize the belief which is also combined with the emotional trigger or particular behavior response.

The body tells the mind it is not safe, and the mind iterates to the body that it is not safe. Through this mutual negative feedback the whole individual is destabilized. According to Transactional Analysis., the adult self uses the game patterns and the script patterns. The organizational activity is the parent self. At a higher level of organization this results in individual complexes. So, levels 7-10 are script-game-racket patterns. We can further speculate that when we experience self as "I/Not-I" we are into the above.

LEVEL 10: BEHAVIORS

This level gives rise to behaviors and the use of the body. Behavior is the interface of the organism with the world. So are the senses, but they are inwardly directed. Behavior is an outwardly directed dynamic. This creates a reaction in the outer world which the senses can perceive and then back to re-evolve belief systems. If this feedback system is flawed or closed, or based on false assumptions, negative beliefs about the self become self-reinforcing. In this manner we create a solid reality that is familiar, predictable, and one with which we can cope. And we find ourselves now back at the surface, having dived deep and discovered experientially the nature of the pure soul and chaotic consciousness.

Changes at these deepest levels effect even the surface layer of behavior in a sure and profound way which unfolds over time. The source of dreams in this model is the most primal or rudimentary level of the psyche. They are a pure spontaneous phenomenon of the brain's experience of itself, turning itself on and off during sleep, sorting and processing input from without and within. They originate in the collective consciousness level as pure consciousness which, as it passes through the layers of self, picks up shapes and plot at all levels to create the dream as we experience it (symbols and plot). Dreams may be the leakage, or extrusion, of this consciousness to the surface level.

Conclusions

This basic, healthy, undifferentiated, collective God-force within percolates into the upper levels of consciousness. And as the dream images filter through each of the levels, they take on shapes which become the images and the plot you see at the surface of the dream. The strength of dream healing is that it gives us the shapes of the dis-ease, the discomforts, the shapes of fears, and of pathology. Playing with just the images of a dream tells us a whole lot about different aspects of the ego, such as how we get along and adapt.

The dream symbols are portals which you can follow back down into deeper levels. Awareness which has made this journey gains a self-transformative power which can be applied to recreating the personality and changing behavior. Then awareness is changed fundamentally. The lost soul is found, and retrieved, and restored. A new sense of wholeness emerges, which is reflected in the personality.

Dream healing takes the sense of self (-awareness) back into symbols to its root levels without interpretation. The interpretation has led in the past to distortion of the information or message from the primal source. The surface level of dream is reflected by plot and symbols. Freud and Jung tended to interpret and intellectualize about dream reality. Fritz Perls approached the dream experientially, with the goal of unifying the elements. Perls remained at the ego levels in his dream work. But now the dreamer can learn directly, experientially that (s)he is all parts of the dream.

References

Gowan, John Curtis, (1975), Trance, Art and Creativity, California State University.

Grof, Stanislaus, Holotropic (2010) Holotropic Breathwork: A New Approach to Self-Exploration and Therapy, SUNY.

Hillman, James, (1977), Revisioning Psychology, William Morrow Paperbacks; First Thus edition.

Hillman, James (1977), "Inquiry into the image," Spring 1977, p. 82 coming 2015:From Types to Images, Uniform Edition Vol. 4, "An Inquiry into Image, Further Notes on Images, Image Sense".

Hillman, James, (1972), Pan and the Nightmare, Spring Pubns,

James, Muriel, and Dorothy Jongeward, (1977), Born to Win, (Addison-Wesley Pub., Philippines).

Jung, Carl, (2010), Dreams: (From Volumes 4, 8, 12, and 16 of the Collected Works of C. G. Jung), Princeton University Press; Reprint edition.

Krippner, Stanley and David Feinstein, (1991), Personal Mythology, St Martins Pr.

Krippner, Stanley, (1991), Dreamtime and Dreamwork, Tarcher.

Miller, Iona and Graywolf Swinney, (1992), Dreamhealing, Asklepia Foundation, Wilderville.

Miller, Iona (1992), "Chaos as the Universal Solvent": Re-Creational Ego Death in Psychedelic Consciousness.

Perls, Fritz, (1992), Gestalt Therapy Verbatim, The Gestalt Journal Press; Revised edition.

Swinney, Graywolf, (1999), Holographic Healing: Dreams, Consciousness Restructuring, Chaos and the Placebo Effect, Asklepia Publishing.

Williams, Strephon K. (1980), Jungian-Senoi Dreamwork Manual, Journey Press; Revised edition.

Article

The Fractal Nature of Active Sleep & Waking Dreams

Iona Miller*

ABSTRACT

Process Oriented Psychology (process work) is a cross-disciplinary approach that facilitates individual and collective change. It offers new ways of working with human difficulties, including emotional disturbances, physical illness or symptoms, relationships, and larger community conflicts. Process work has its philosophical roots in Consciousness Studies, Jungian Psychology, Chinese Taoism, and Shamanism. The hypothesis of Harvard biopsychiatrist, C.M. Anderson (1998) provides some enticing substantiation for psychophysical restructuring in shamanic journeys and other process work in therapy. His work is centered around the psychotropic and oneiric aspects of the shamanic entheogen iboga, used by the Bwiti tribe of Africa, and employed therapeutically by Harold Lotsoff and Dr. Robert Goutarel for the elimination of chemical dependency and Post-traumatic Stress Disorder (PTSD). Other research supports some of these conclusions.

Drug-free shamanism shares many common features with this therapy: ego-death, initiatory waking-dream journeys, shamanic guidance, quick life review, psychedelic states of consciousness, spontaneous fetal regression and perinatal states, journeys to "the land of the dead," and the induction and facilitation of Self-organized Critical States (SOCs) which result in restructuring of fundamental neural patterns. Stress or abuse in early life leads to hemispheric asymmetry which is implicated in a wide variety of disorders. Process work allows us to "exercise" unused pathways and reinstate hemispheric synchronization. Disorders of under- and over-arousal can be dynamically balanced, reinstituting organismic equilibration. The fractal nature of REM allows us to process and restructure old emotional patterns, at the sensorimotor root by reviving neural plasticity. CRP (Consciousness Restructuring Process) facilitates the Self-Organized Critical state (SOCs) which initiates cascading therapeutic reactions which are robust and persist over time.

Key Words: emotional memory, amygdala, REM, NDEs, psychedelic states, self-similar stochastic fields, fractal nature, shamanic healing, dreams, psychotherapy, spirituality, alternative health.

Introduction

To many death seems to be a brutal and meaningless end to a short and meaningless existence. So it looks, if seen from the surface and from the darkness. But when we penetrate the depths of the soul and when we try to understand its mysterious life, we shall discern that death is not a meaningless end, the mere vanishing into nothingness--it is an accomplishment, a ripe fruit on

*Correspondence: Iona Miller, Independent Researcher. Email: iona_m@yahoo.com Note: This work was completed in 2001.

the tree of life. Nor is death an abrupt extinction, but a goal that has been unconsciously lived and worked for during half a lifetime.

In the youthful expansion of our life we think of it as an ever-increasing river, and this conviction accompanies us often far beyond the noonday of our existence. But if we listen to the quieter voices of our deeper nature we become aware of the fact that soon after the middle of our life the soul begins its secret work, getting ready for the departure. Out of the turmoil and error of our life the one precious flower of the spirit begins to unfold, the four petaled flower of the immortal light, and even if our mortal consciousness should not be aware of its secret operation, it nevertheless does its secret work of purification. ~Carl Jung, The Symbolic Life, Page 757

Carl Jung, Carlos Castenada, Elizabeth Kubler-Ross and others have suggested we keep the personification of our own death close to our waking consciousness as an ultimate advisor and guide to the labyrinth of inner and out life. Some psychoactive substances as well as psychotherapeutic process work can plunge us into experiential simulations of death and rebirth, deconditioning and restructuring consciousness at the primordial level. Breakdown can lead to breakthrough, spontaneously or in a structured therapeutic or shamanic setting.

Anderson (1998) hypothesizes that the ingestion of iboga (or its alkaloid ibogaine, ibogaine hydrochloride) in a shamanic or psychotherapeutic setting induces a critical oneiric, or dream-inducing state, in which fractal time patterns of phasic events similar to those existing during fetal rapid eye movement (REM) or Active sleep are recreated in the adult.

This dynamically destabilizes the functional connectivity of the brainstem and its habitual interactions with the bihemispheric temporal lobe structures such as the amygdala, creating a functional *state of plasticity* in these areas which facilitates the reintegration of traumatic memories. Psychopathological interhemispheric dynamics are altered, dissipating old behavioral attitudes and patterns in a simulated near-death experience (NDE).

This psychotherapeutic oneiric state is similar to the complex behavioral states of REM sleep and attentional orienting in that they all share the signature of the self-organized critical state. Observed similarities between the neurophysiology of the REM state and that induced by selective psychedelic drugs such as LSD or psilocybin further substantiate this hypothesis. The dream-like quality of these journeys and the emergence of discrete states of consciousness (d-SOC) exhibiting particular imagery is reported by LSD-researchers (Tart, 1969, and Grof, 1988).

Anderson highlights the common ground of REM sleep, orienting, and psychedelic states. Drug-free shamanic journeys using Consciousness Restructuring Process (CRP) can also enable us to "return to infancy and birth"- to the life in the womb - by returning us to the uterine condition (Grof's parinatal matrices). It facilitates a condition in any case very close to life in "the land of dead" (realm of archetypes and NDEs) and so restores us to our own integrity -- our pristine condition.

In the Bwiti and Fang rites, great importance is given to the *retreat* and the *confession* which precedes the initiation. Both aspects, as well as *purification* are essential features of the

Consciousness Restructuring Process (Swinney and Miller, 1992; Swinney, 1999). Ibogaine therapy and CRP share the feature that the imaginal material elicited is easy to manipulate either on the journeyer's initiative or by the mentor.

The therapist can stop to contemplate a scene, go back, explore an alternative in a given sequence, bring a previous scene back to life, etc. The ease with which the events in treatment can be manipulated and the fact that the experience can be directed to dynamic areas in a flowing and self-emergent way, is probably one reason for its psychotherapeutic success.

All the while the journeyer reports waking dream sequences without loss of consciousness or any illusions of formal deterioration of thought. They report feeling suspended in the stream of time, and being shown where their problem is. The key feature shared by both modalities is temporary destructuring of the ego, followed by its restructuring, and symbolic rebirth.

Michel Jouvet and Sir Frances Crick have assessed the role of dreams in the programming and de-programming of basic behavior patterns, resulting in a new individuation of the human brain. They consider PGO waves to be the principal coding tool that acts at the cortical level in recording the genetic and epigenetic acquisitions necessary for the individuation of the human brain.

Ponto-geniculo-occipital waves or PGO waves are phasic field potentials. These waves can be recorded from the pons, the lateral geniculate nucleus (LGN), and the occipital cortex regions of the brain, where these waveforms originate. The waves begin as electrical pulses from the pons, then move to the lateral geniculate nucleus residing in the thalamus, and then finally end up in the primary visual cortex of the occipital lobe. The appearances of these waves are most prominent in the period right before rapid eye movement sleep (or REM sleep), and are theorized to be intricately involved with eye movement of both wake and sleep cycles in many different animals. (Wikipedia)

PGO waves branch out into a network leading the phasic electrical signal toward the lateral geniculate nucleus and the occipital lobe. This modulates executive neurons, triggering and transfer neurons, aminergic-cholinergic neurons, nitroergic and GABAergic neurons, and more. There is a relationship between plastic epigenetic gene expression, DNA and dreams – perhaps by enhancing DNA repair capacity by affecting genes involved in DNA damage responsive pathways.

Research (Nelsen, 1983) suggests that, "PGO burst cells are output elements in the PGO wave-generation system ad that PGO waves convey eye movement information to the sensory visual system in REM sleep. They also may have a role in the production of saccade-related waves in the visual system during wakefulness."

Metaphorically, this "software-writer" aspect of the self allows one to reconfigure the genetic and cultural programming much like changing the *config.sys* file of a computer. REM then reboots the consciousness patterns with the *autoexec.bat* file for habits, needs, and the manner in which one approaches life (Miller, 1993).

In addition, through "chaotic" activation mechanisms, the PGO waves eliminate from certain types of neuronal networks an informational overload linked to pathological behavior. This is what Debru (1990) calls "*cleaning out the neuronal circuitry.*" Huxley made a similar comment about "cleansing the doors of perception." Apparently REM sleep undergirds a sorting out process among the "residues" stirred up by the PGO wave sleep pattern and disposes of these residues during dreaming.

The principal difference between dreams and hallucinations resides in the way in which the stages of wakefulness are organized, with the suppression of REM sleep and the intrusion of PGO waves in the arousal (waking) stage and in NREM (or slow) sleep. The new organization becomes: waking (arousal) stage, stage of PGO waves, hallucination stage, sleep stage, and it appears possible that hallucinatory manifestations, the waking dream, eliminate "residues" stirred up by the PGO wave pattern in the absence of REM sleep.

These visions are analogous to those at the approach of death, or what are called near death experiences (NDEs), which appear as a tunneling or journey toward the hippocampus as various brain areas discharge and go offline. They are the same as those termed normative visions. Such visions include the characteristics of two phases of NDEs (Sabom, 1982):

The Autoscopic phase includes *1) subjective feeling of being dead; 2) peace and well-being; 3) disembodiment; 4) visions of material objects and events.*

The Transcendental phase includes *5) tunnel or dark zone; 6) evaluation of one's past life; 7) light; 8) access to a transcendental world, entering in light; 9) encounter with other beings; 10) return to life.*

Turn Off Your Mind, Relax & Float Downstream; It is Not Dying

Jansen (2000) has a ketamine-based model of near-death experience, claiming a central role for NMDA receptors. His hypothesis links most of the neurobiological and psychological theories (hypoxia, a peptide flood, temporal lobe electrical abnormalities, regression in the service of the ego, reactivation of birth memories, sensory deprivation etc.) rather than being an alternative to them. Super-psi theories have created an impasse in explaining survival evidence. But, he summarizes without spiritualizing or sensationalizing:

Near-death experiences (NDE's) can be reproduced by ketamine via blockade of receptors in the brain (the N-methyl-D-aspartate, NMDA receptors) for the neurotransmitter glutamate. Conditions which precipitate NDE's (hypoxia, ischaemia, hypoglycaemia, temporal lobe epilepsy etc.) have been shown to release a flood of glutamate, overactivating NMDA receptors resulting in neuro ('excito') toxicity. Ketamine prevents this neurotoxicity. There are substances in the brain which bind to the same receptor site as ketamine. Conditions which trigger a glutamate flood may also trigger a flood of neuroprotective agents which bind to NMDA receptors to protect cells, leading to an altered state of consciousness like that produced by ketamine.

Author of the pioneering book, *Neurotheology*, Laurence O. McKinney (1994) calls his "eternity first/death later" "back to the brainstem" approach the literal unraveling of the brain in normal brain death.

Oscillatory activity in the brain is widely seen at different levels of observation and is thought to play a key role in processing neural information. The brain has to make use of analog representation because of the speed limitations and biological fragility of its neuronal structures, and the need to combine many parallel streams over short periods of time. The correlation between neuronal structure and brain capacity is evident throughout species. Research correlates evidence that neurons employ network oscillations to achieve analog representation.

From a brain science point of view, I started by checking the parallels in "creation" and "going to heaven" traditions from all the religions, and, they seemed to be addressing the same experience - which led to eternity as a state of no chronology. This changed me into the basis of "where did we come from, why are we here, where are we going" as simple artifacts of a chronological consciousness that only gains traction as the prefrontal area starts arranging memory when the major brain growth is over. It also explains morality- the moment you can imagine a future retribution you have a reason to behave in a human society, End of story. (Private correspondence, 2011)

"In the dying brain, visual cortex disinhibits concurrent with last voltage spikes, and there's your bright white - then it rolls down the spectrum to black. Prefrontal lobes go first, providing the no-time horizon and brain becomes less and less complex overall, sending any consciousness back into infantile and fetal 'thought'; we blank out in all encompassing eternity as far as we're concerned."

"The world we're living in is actually living in us - less than a heartbeat away from reality" ... "but this is how we each, as an individual, experience our own creation, maturation, the time of our lives, and the release into eternity." "In a tenth of a second it's the world within that we experience, and it has it's own birth ... and it's own death." ... "not only is this universe within - but in the time it takes to assemble our universe another universe could have been born - nice thought, from a Hindu perspective." (McKinney)

Todd Murphy describes phenomena of the cingulated gyrus and angular gyrus in Out-Of-Body experiences: "The angular gyrus is thought to play a role in the way the brain analyzes sensory information that allow us to perceive our bodies. When it misfires, they suggest, the result can be a sense of floating, and seeing the world from outside of the body. The findings were published . the respected science journal "Nature". The angular gyrus is involved in our perception of our own bodies."

V. S. Ramachandran, director of the Center for Brain and Cognition at the University of California, San Diego, describes how the angular gyrus is probably at least partly responsible for the human ability for comprehending metaphor. Patients with damaged angular gyrus, Ramachandran said, "often came up with elaborate, even ingenious interpretations – that were completely off the mark." The implication is that OBEs are metaphors that can be taken literally and so 'feed' the concept of being able to 'escape' the body.

Carhart-Harris et al., have shown the neural correlates of the psychedelic state as determined by fMRI studies with psilocybin. This psychedelic activates the brain hubs in the "default mode network (DMN) — a network of brain regions that becomes active when you allow your mind to wander. So what apparently happened is that the subjects were able to achieve an unconstrained style of cognition — in other words, totally trip out without the usual reality constraints."

"These results may have implications beyond explaining how psilocybin works in the brain by implying that the DMN is crucial for the maintenance of cognitive integration and constraint under normal conditions," the researchers said. "This finding is consistent with Aldous Huxley's 'reducing valve' metaphor and Karl Friston's 'free-energy principle,' which propose that the mind/brain works to constrain its experience of the world."

They also discovered that the psychotropic improved people's ability to access personal memories and related emotions, which could be helpful in psychotherapy.

In order to feel that time, fear, and self-consciousness have dissolved, certain brain circuits must be interrupted. This includes damping activity in the fear-registering amygdala, which monitors the environment for threats. Parietal-lobe circuits that give us a sense of physical orientation and a distinction between self and world must quiet down. When the parietal lobes quiet down, a person first feels detachment from the tyranny of the perceptions, then an expansive oneness with the universe of cosmic unity.

The orientation area requires sensory input to do its calculations. Intense meditation blocks the brain from forming a distinction between self and world. Frontal and temporal-lobe circuits, which mark time and generate self-awareness, must disengage. When this happens, self-awareness briefly drops out and we feel like our boundaries dissolve.

Time distortion starts the personal escape from time, sign of an attempt to escape from the cocoon. Then the inner marriage between the personal and non-personal aspects of the psyche is consummated. Psychic conflicts are transcended, leaving a whole, complete being.

When the orientation area is deprived of neuronal input by gating from the hippocampus, sense of self expands. With no preferred position or direction in space, the local self dissolves in omnidirectional expansion. If one remains motionless, there is no external reference signal to orient in 3-space and no reason for this portion of the brain to activate. Continued meditation can over-drive certain other brain areas and seemingly transport us to another universe.

For a mystical experience to occur perceptions narrow and brain regions that orient us in space and mark the distinction between self and world must go quiet. We give up the past and give up the future -- we give up our defenses. All feelings cease as the self merges with the numinous element. The mind becomes tranquil, withdraws itself from all sides, becoming firmly established in the supreme Reality.

The most immediate experience is that of always having been and being forever. The three illusions -- space, time and personality -- are obliterated in cosmic consciousness, as the soul completes its journey to its spiritual home. Human consciousness is eliminated, having been reabsorbed into the primordial essence. All becomes All without differentiation.

Endogenous Hallucinogens

Strassman (1990) has suggested that the *pineal gland* is a possible source of endogenous hallucinogens and this gland is also associated with sleep cycle rhythms, and traditionally with mystical states of consciousness. Besides the production of melatonin, the pineal may synthesize endogenous hallucinogens in response to certain psychophysical states, and raise serotonin levels in the brain.

These hallucinogens may belong to the tryptamine or beta-carboline family of compounds. One compound (6-methoxy-1,2,3,4-tetra-hydro-beta-carboline) has been proposed as the producer of rapid eye movement sleep. It is concentrated in the retinae of mammals which may be related to its visual effects.

There are several ways in which either psychoactive tryptamines and/or beta-carbolines may be produced within the central nervous system (and possibly within the pineal) from precursors and enzymes that are known to exist in human beings. In addition, nerve fibers leave the pineal and make synaptic connections with other brain sites through traditional nerve-to-nerve connections, not just through endocrine secretions.

Serotonin or tryptamine levels are higher in the pineal than any other organ in the brain. 5-methoxy-tryptamine is a precursor with hallucinogenic properties which has a high affinity for the serotonin type-3 receptor. Gucchait (1976) has demonstrated that the human pineal contains an enzyme capable of synthesizing both DMT and bufotenine-like chemistry. These compounds are prime candidates for endogenous "schizotoxins," and their production may be related to stress and/or trauma, and has been implicated in the etiology of schizophrenia.

Strassman notes that both the embryological rudiments of the pineal gland and the differentiated gonads of both male and female appear at 49 days. Melatonin is a time-keeper for gonadal maturation and/or competence so the pineal is implicated again. He suggests this may be the ontological source of the tension between sexual and spiritual energies. The pineal gland, as source of both psychedelic compounds and the gonads, source of physical immortality, may work in concert (or oppositon) in the individual's development through time.

Stress-related hormones are implicated in pineal activation to activate normally latent synthetic pathways, creating tryptamine and/or beta-carboline hallucinogens. When we face stress, potential death or in meditative reveries, we "tune back" into the birth experience, the most well-developed motif of such experiences. Perinatal themes and embedded memories re-emerge.

Those with Caesarean deliveries report greater difficulty in attaining transcendent states of breakthrough and release during drug-induced states. Maybe less fetal (or maternal) hallucinogens were released at the time of birth. They may not, according to Strassman, have a strong enough "*template of experience*" to fall back on, to be familiar enough with to let go without fear of total annihilation, because lesser amounts of pineal hallucinogens were produced during their births.

The pineal may be modulated in its activity by meditative practices, to elicit a finely-tuned standing wave through resonance effects and other techniques. It creates the induction of a dynamic, yet unmoving, quality of experience. Such harmonization resynchronizes both hemispheres of the brain.

Dysynchrony is implicated in a variety of disorders. Such a standing wave in consciousness can induce resonance in the pineal using electric, magnetic or sound energy, and may result in a chain of synergetic activity resulting in the production and release of hallucinogenic compounds.

Thus, the pineal may be the physical representation of an attractor, or "lightning rod" of consciousness. Pineal function may profoundly affect consciousness at the time of birth, death, near-death experiences, and during unusual psychophysical states such as CRP Journeys or meditation.

Hemispheric Asymmetry

Stress or abuse in early life induces abnormal hemispheric functional asymmetries, disrupting REM sleep and predisposing individuals to addictive and self-defeating behaviors resulting from impaired interhemispheric integration.

The profound and persistent neural and psychological changes and vulnerabilities induced by early trauma are revealed with EEG during recall of past trauma. Further, abnormal hemispheric EEG coherence is associated with reductions in the size of the corpus callosum, the bridge between hemispheres. It has been proposed (Schiffer, 1997, 1998) that we all have two minds or personalities, one in each hemisphere. They are like conjoined Siamese twins, and their disharmonious struggles for dominance result in a range of personality disorders.

The emotional mind, largely in the right hemisphere, may be damaged from trauma or abuse, and sabotage the critical mind. We always know what we *should do*, yet we do what we want, even when it leads to self-defeating consequences. The hemispheres are meant to work in concert with one another. Debilitating emotional disregulation results from hemispheric disharmony and dysfunctions in the arousal system toward hypo- or hyper-tonic states.

The left hemisphere (with the frontal lobe) manages tonic activation for the conduct of intellectual and motor tasks, and maintenance of vigilance over time. It is mediated by the neuromodulators dopamine and acetylcholine. The right hemisphere, in contrast, manages phasic arousal to maintain the sensory system in readiness to receive and process new inputs from any source.

This system is cued by norepinepherine and serotonin. An increasing number of disorders are being assigned to one hemisphere or the other. Neurofeedback uses a number of protocols for the correction of alpha asymmetry which is implicated in PTSD, depression, ADD, addiction,

OCD, anxiety, and a host of other dysfunctions. Human minds transcend the hardwiring of the brain; dynamic brain plasticity is one mechanism by which they do so.

Physicist David Bohm called a noun a "slow" verb, and we humans are constantly in dynamic psychophysiological motion even if we consider our personalities to be "fixed." In ordinary waking consciousness the two hemispheres--linear left-brain and holistic right brain--exhibit uncoordinated, randomly diverging wavepatterns detected by the electroencephalograph.

When we enter a meditative state, these patterns tend to become synchronized, and in deep meditation they exhibit nearly identical patterns. In deep meditation not only do the left and right brains of one subject synchronize, they can resonantely entrain with others in the vicinity, as paired subjects synchronize.

The "*music of the spheres*"--a brain symphony, plays when hemispheric synchronization occurs, and both lobes function in concert. A hemispherically-balanced mind is an "*open-mind.*" Lateral specialization of the cortex into two distinct and complementary modes of consciousness reveals that the left brain excels at verbal skill, linear thought, abstraction, rationality, and analytical thought, whereas the right is more nonverbal, synthetic, global, diffuse, metaphorical, dreamy, imaginal, perceives gestalts, and helps us with visual construction and spatial orientation.

The right hemisphere gives us non-linear leaps of intuition and insight--the 'a-ha' experience. It is subjective, relational, holistic and time-free. The right brain is the dreamer and artist. It gives us the startling perceptual experiences produced by drug-free experiential therapy or consciousness journeys, such as mind trips, bursts of ecstatic feelings or sequentially logical thoughts, insight, followed by a cognitive evaluation period. There is a possible gating-mechanism which seems to occur in CRP when someone recalls a dream or symptom.

The influx of perceptions produced by increased attentiveness and sensitization to sensory stimuli may overwhelm the systematic sequential processing of the language hemisphere and invoke the analogical integrative mode of the right hemisphere to consolidate the perceptual flood. Yet, a whole brain is better than either alone. Collective neuronal activity is modulated by rhythmicity, and this is what is detected with the EEG. Neuronal populations coalesce to collective firing when stimulated or processing. Then they desynchronize back to raw signal.

Desynchronization results from the superposition of many rhythmic generators of different frequencies, each ebbing and flowing from rhythmicity to desynchronization. Rhythmicity regulates the entire spectrum of activation and arousal in the bio-electrical domain by a process called kindling. The process breaks down when synchronization or desynchronization of specific frequencies persists or is dis-regulated, decoupled from the demands of the moment. Self-recovery reinstitutes harmonization of formerly hyper- or hypo- arousal states of the person.

Disorders of underarousal include unipolar or reactive depression, inattentive-type ADD, chronic pain, and insomnia. Overarousal includes anxiety disorders, sleep onset problems, hypervigilance, impulsive ADD, anger/aggression, agitated depression, chronic nerve pain, and spasticity. Some forms of anxiety and depression involve both under- and over-arousal. Some instabilities arise autonomously from the CNS while others require an external trigger for initiation.

Early trauma creates vulnerabilities to both. Over- and under-arousal indicate changes in sympathetic and parasympathetic arousal, called ergotropic and trophotropic shifts. The ergotropic, or energetic shift is characterized by a tendency toward higher sensory acuity, external focus, sympathetic arousal, high motor setpoint, etc. The trophotropic or tranquilized state is a tendency toward an inward focus, less alertness, reduced sensory acuity, a shift toward vegetative functions, and reduced motor readiness.

By stimulating the neglected neural circuitry, new pathways are created, improving equilibrium and long-term change. Ergotropic and trophotropic shifts are mutually inhibitory. To enhance one is to suppress the other. Their dynamic balancing results essentially in a 'tuning' of the nervous system. We can conclude that CRP helps us switch from left brain to right brain dominance during the journeys, and that it facilitates neuronal restructuring which reinstitutes hemispheric synchrony and the wider distribution and amplitude of alpha and theta waves, reunifying the whole brain, (Miller and Swinney, 2001).

The brain's intrinsic bias toward homeostasis dictates that any process which evokes a brain response away from its then-prevailing equilibrium state will set in train forces to restore the original state. Thus promoting arousal by focusing on fear and pain will first tend to produce a shift, and on the other hand, set in motion a compensatory mechanism by which the brain restores its opposite and enters a calm period. Hence, even dis-equilibration can bring out improved equilibrium maintenance as a long-term consequence. That dis-equilibration is the introduction of deterministic chaos.

CRP offers a regulatory challenge to the dynamic system by taking the brain momentarily out of its prevailing equilibrium. The brain responds positively to therapeutic disequilibration of the nervous system by long-term adaptation. Ultimately, it doesn't matter whether the disequilibration occurs in one direction (breakthrough therapy) or the other (meditation). Improved regulatory function eventuates in either case. CRP is a method by which the functional or systems level of the brain-mind is addressed.

The functional process can be immediately altered and new patterns of behavior and perception facilitated. CRP leads to long-term dendritic re-programming and/or regrowth, and significant reorganization. It "exercises" neglected pathways. Because it works so quickly, the mechanisms involved in stabilization must lie principally in the functional rather than structural level, at least initially.

Functional plasticity is undoubtedly mediated by altering synaptic coupling strengths through the generation or attrition of receptor sites, and the alteration of neurotransmitter chemistry through changes in neuronal gene expression, (Rossi, 2000). The functional plasticity of neuromodulator systems clearly exists on all behaviorally relevant timescales. The dynamic range of neuromodulator plasticity or flexibility can be increased where it is deficient, and stabilized when it is unstable by self-organizing restructuring.

CRP simultaneously exercises neural mechanisms which control the fundamental functions of arousal, attention and affect managed by the central nervous system (CNS). It works through a complex web of inhibitory and excitatory feedback networks. Both functional and structural changes are implemented.

The left hemisphere aspects of depression and anxiety may have to do with anticipatory activity, planning, ruminating, perseverating, and worrying. The right hemisphere, in contrast, harbors the non-rational, more catastrophic aspects of depression and anxiety, (for example, including fear, panic, agitated depression, and suicidality). There are hemispheric specificities for cognitive function, anxiety, depression, pain, sleep disorders, eating disorders, endocrine and immune system disorders (Othmer et al).

Skills training for the right brain requires tasks that the left, rational mode gets easily bored with--things it either can't or won't do. In right brain mode we become unaware of the passage of time, are alert but relaxed, excited but calm; it is relief to relinquish rationality temporarily--this is the source of the age-old craving for self-induced altered states of consciousness which bring meaningful satisfaction, (Edwards, 1999).

The hemispheres affect motor abilities also, since the left hemisphere is hard-wired to the right side of the body and vice versa, with the exception of a small percentage of left-handed people. Left-handers are generally less lateralized; they tend to process language and spatial information in both hemispheres. This creates potential problems or conflict such as dyslexia, but it also can lead to superior mental abilities. The hemispheres are virtually identical in appearance, though not in function.

The corpus collosum, which connects the two halves is larger and has more connections in females than males. If severed, the two halves of the brain continue to function independently with differing perceptions and agendas. In fact, both halves are involved in higher cognitive functioning, each half specialized in complementary fashion for different modes of thinking. Their harmonization gives us the sense of being one person--a unified being. Research shows indications that the modes of processing tend to interfere with each other, preventing optimal performance. This may be a rationale for the evolutionary development of asymmetry, as a means of keeping the two different modes of processing discrete.

The corpus collosum is the gating system between hemispheres. States, rhythms, and resonances function beyond the realm of hard-wired neuronal connections. In split brain experiments, showing a subject two different items or photos for each side, a subject may incorrectly identify an item with the left hemisphere, but the right lacking verbal skills to correct it, will "speak" through the body such as by shaking the head "No." The other way of knowing is the source of "gut feelings." While at the analytical level, the subject will conjecture as to why that is happening. Each half has its own way of knowing about our being and perceiving external reality.

Each of us is of two minds, mediated by the connecting cables of the nerve fibers in the corpus callosum. Yet these hemispheres can work together in a number of cooperative ways. There are two discrete ways of knowing. The main distinctions in hemispheric processing are among thinking and feeling; intellect and intuition; objective analysis and subjective insight. The history of science is full of anecdotes about researchers who have a dream or intuitive hunch where a metaphor spontaneously presents as a creative solution.

The two modes of conscious can each be the leader or the follower. They may also conflict, one half trying to do what the other knows it can do "better." Each has its own way of keeping knowledge from the other hemisphere, and this is especially true when it comes to memories and patterns locked in from trauma and abuse.

The Fractal Nature of REM and Metaphor

Patterns of bunching or clustering typically reveal an underlying fractal organization inherent in nature. Living and nonliving matter is organized into complex recursively nested patterns over multiple levels of space or time, or spacetime (the holomovement). Patterns termed fractal or self-similar are recurrently irregular in space or time, with themes repeated like the layers of an onion at different levels or scales. Fractals are nested patterns which are robust phenomena which persist at all levels of observations.

Complex dynamics and fractals are ubiquitous in human psychophysiology and feedback loops. They can be seen in the bunching or clustering in the opening and closing events of ion channels, quantal release of neurotransmitters, or the spontaneous firing patterns of neurons, heartbeats, or breaths from the fetus to the adult. As we have not failed to notice in CRP, Anderson reiterates that fractal concepts provide an essential point of view for understanding brain/mind.

There are fascinating connections among REM sleep, attentional (orienting) and psychedelic (or mind-manifesting, mind-expanding) states. As in the case of his ibogaine hypothesis, CRP therapy intervenes in the cortical-amygdaloid-brainstem loops in hemispheric disharmony. To make a stronger statement, there are neurophysiological similarities between REM sleep, orienting and transcendent states which have implications for any dream-inducing operators, and CRP is one such drug-free operator.

In CRP as in life, these are more than conceptual or metaphorical links. They emerge directly from our sensorimotor fleshly nature and bear on our psychological and philosophical nature as well. In many ways, we agree with Lakoff's (1988) definition of metaphor as a schema, *"a unifying framework that links a conceptual representation to its sensory and experiential ground."* His central thesis that metaphors facilitate thought by providing an experiential framework in which new information may be accommodated, forming a cognitive map, a web of concepts rooted directly in physical experiences, and our relation to the external world.

This cognitive topology is a mechanism we use to impose structure on space. Another core idea of Lakoff's is that clusters of metaphors describe experiences better than any single metaphor can. Because of the ubiquitous nature of clustering in fractal systems, we suspect that complex dynamics are at work in our inherent experience of ourselves and our reality through epistemological metaphors. Thus metaphor is much more than a superficial phenomenon of language--not a means of expression as much as a means of apprehension, which shapes our thoughts and judgments, and structures our language.

Our perceptual experience is rooted in a few key conceptual categories which Lakoff has defined:

1) *Thought is embodied:* it grows out of bodily experience and makes sense in terms of it; we are grounded in perception, body movement, and our physical and social character.

2) *Thought is imaginative:* it unfolds spontaneously in terms of metaphor and psychophysical imagery, which is much more than literal. This imaginal capacity is also embodied indirectly in metaphors and images based on experience, particularly bodily experience.

3) *Thought has gestalt properties:* this is neither a structuralist nor functionalist perspective, but one of radical nondualism. Self-organization with emergent properties is a descriptor of dynamic processes with a fractal blueprint.

4) *Thought has an ecological structure:* nature follows the path of least resistance in its webwork of synergetic interaction, and the ecology of systems depends on the overall structure which is in constant dynamic motion.

Thus thought is much more than the mechanical manipulation of abstract or abstracted symbols. Symbols do not require interpretation, but arise as emergent properties with inherent "meaning." Thought is strongly rooted in the neurology of the brain, in orientation-sensitive cells, and center-surround receptive fields, the interface of the part with the whole. The sensory-motor system is fundamental in this orienting, as is metaphor which builds our neural maps. This allows sensory-motor structures to play a role in even abstract reasoning.

Because of the link in this perspective between sensory-motor experience, orienting behavior, and metaphorical apprehension, Lakoff's notions can be strongly linked to Anderson's vision of interhemispheric reintegration with its emergent neural plasticity, reorientation, restructuring, and transcendent capacities, all of which he links to dream-inducing REM sleep, with its fractal patterning.

The same applies to CRP. Our automatically called-up metaphorical perception continues in our sleep life, and demonstrates that imaginal life is fundamental to our existential perspective, rather than an artifact of trying to describe our experiences. This harmonizes with CRP which empirically notices that the root of our existential self-image lies in the sensorimotor root.

The onset of REM suppresses a complex network of serotonergic neuron cell body groups (the S-Net) in the brainstem (the RAS and dorsal raphe nucleus, DRN), and psychedelics mimic this action, as we suspect endogenous hallucinogens do. During both experiences there is significant depression of the electrical activity of the brain's serotonin-containing neurons.

According to Anderson, therefore, "*the change in raphe unit activity seen spontaneously across the sleep-waking cycle may be the key to understanding altered states of consciousness.*" The key serotonin system may function in a manner appropriate to a different behavioral state, such as REM sleep while the subject is still awake--awake yet actively dreaming. The PGO waves which induce phasic eye movements of REM are readily observable via EEG and by inference through direct observation. It doesn't take a drug to induce the state.

In the therapeutic setting just the simple suggestion of its possiblity, permission, and a positive expectation for the state allow altered states to emerge and unfold. This phenomenon is well-documented in the literature of clinical hypnosis. In neurological terms, the fractal patterns of reorganized S-Net unit activity are allowed to emerge through an autopoietic process, in concert with dynamic changes in other brainstem and forebrain areas. Many of the unpredictable patterns of oneiric (dream-induced) behavior can occur.

1/f Fractal Patterns and Time-Binding in REM, Orienting & Transpersonal States

Anderson proposes and delineates a connection between REM sleep and attentional states, such as orienting. He contends that mammals in general are in a state of virtually continual orienting during REM sleep, and relates this to implications about discrete states of consciousness. He cites experiments where similar amplitude PGO waves are evoked in cats during normal orienting responses to loud sounds and during normal REM sleep when external behavioral orienting is absent.

PGO waves suppress the 5-HT or serotonin cycle. The 1/f fractal patterns typically seen in cat brains during orienting to birds, for example, is linked to REM sleep, but not other types of sleep and quiet wakefulness. This fractal pattern is diminished by serotonergic antagonists. Wakefulness and slow wave sleep is not conducive to the 1/f state. But the fractal pattern becomes very active at the offset of these states, including REM, orienting, and transcendence. Anderson concludes and defends that this suggests that REM sleep is, at least in terms of 5-HT systems, a prolonged orienting response.

And we know from Lakoff, that humans tend to orient in terms of metaphors, and cognitive maps based on sensorimotor experiential metaphors. This creates unique 1/f fractal patterns of activities in time, and in fact, is a mechanism of time-binding. Imaginal or physical movement in three-space can only take place in the 4th dimension of time, even in virtual reality.

For example, a metaphor, such as "*love is a journey*" implies a timeline where there is an esoteric, ill-defined something in front of us, a history which lies behind, and a plethora of tangential "*roads-not-taken*" which veer off at all-possible vectors. This phase space functions as a cognitive map of our motion through time. Induced dream-like states have many of the behavioral and neurophysiological markers of REM sleep without atonia, or sleep paralysis.

Anderson suggests this may result from sudden massive destabilization of the normal behavioral-states-rhythmicity of the tonically firing S-Net, forebrain and reticular formation. This effects a complex dynamic network of interdependent dopaminergic, norandrenergic, opiod, cholinergic, and NMDA receptors and systems. Any means that switches dominant activity to the right hemisphere will create a shift from normal attention to a dream-like state, much like switching a channel. It can also jolt the S-Net and Reticular Formation (RF) into a unique state. This unique state, according to Anderson, "*may require a return to activity patterns more characteristic of fetal ontogeny to reinstate normal functional organization.*"

In a sense, following treatment, the Reticular Formation and associated brains regions are functionally "born again." To see how CRP may work through a "fetal REM-like" state in humans, it is necessary to review the important role of REM sleep during development. We must also bear in mind Anderson's finding on the fractal-in-time nature of fetal REM sleep phasic processes and their disruption by early stress. He suggests visualizing these unique states with 1/f patterns, fractal patterns whose complex cores are attractor-based.

Fractals are objects in space or fluctuations in time that possess a form of self-similarity. Fragments of the object or sequence can be made to match the whole by shifting and stretching. Fractals can be exact or statistical copies of the whole, nested self-similar patterns and structures which emerge from this patterning. Fragments of natural fractals are only statistically related to the whole. Only purely mathematical fractals can be exact copies.

In nature, self-similar clusters have smaller clusters within larger clusters of clusters. Clustering patterns, or bursts within bursts, are a universal characteristic of spontaneous behavior in living systems of cells, neurons and the early motility of embryos.

Biological systems thrive and grow via self-organized fractal bursts patterns. Self-similar bursts-within-burst patterns are ubiquitous, and can be seen in ion channel current fluctuations, neurotransmitter release, neuronal firing patterns, the searching and orienting patterns of animals, and in human judgment and decision making. Bifurcations or catastrophic state-changes occur when the organism or system reaches the self-organized critical state.

The Self-Organized Critical State

Complexity theory is applicable to any level of biological or psychophysical description. Its fundamental concept is that of the self-organized critical (SOC) state, which describes how complex spatially distributed entities, such as our organismic feedback networks, can interact across many time and space scales.

This describes the 1/f fractal patterns and their seemingly random yet deterministic perturbations. The SOC leads directly to cascading restructuring of the pattern, system, or structure. The rule of thumb is that "*changes that are twice as big, occur half as often.*" A dynamic system in the critical state produces chain reactions of all sizes and durations.

The SOC is critical in generating patterns by which psychological processes unfold through time. It is also fundamental to the emergent, persistent structures which arise as consequences of this unfolding. They show up at all psychophysical levels, including cognitive, behavioral, non-linear, quantal, sociological, phenomenological, and transpersonal scales.

The Self-Organized Critical State appears in REM, orienting and transpersonal experiences. "*Under the influence*" of the SOC we virtually cannot distinguish our inside from outside; in a sense it turns ourselves "*inside out.*" The imaginal experiences are "real" in that they carry real consequences in the ordinary world. REM, orienting and mystic journeys all share a common critical state that exists throughout the brain and brainstem--patterns of interspike intervals.

Anderson proposes, *"PGO spikes and other phasic activity during these states, are analogous to sand slides or traffic jams of all sizes [ref. chaos theory] representing critical fluctuations in neural activity and connectivity. The SOC state during the orienting response, may facilitate rapid functional brain reorganization in response to the qualities of the eliciting stimulus. The critical connectivity that exists during these states may primarily involve orienting synergies (among ocular, neck and facial motorneurons). PGO waves may link this critical brainstem centered connectivity with limbic and cortical structures such as the amygdala and temporal lobes."*

PGO spike density increases as tonic REM sleep begins. Therefore REM may be a dense, coalescence cluster of PGO activity. From the fractal point of view, REM sleep is a kind of fractal of PGO bursts. With eyes closed, during the oneiric state, PGO-like spikes among amygdaloid and brainstem sites could generate and direct waking dream sequences, according to Anderson. This seems a plausible mechanism for phenomenon observed in CRP Journeys, as well.

Orienting of attentional states is directed inward in the virtual environment, rather than acted out externally. Perhaps this deep focus allows the S-Net pauses to allow sensory processing, and possible motor system functional reorganization. For example, Anderson offers the following in regard to ibogaine's reprogramming capacity: *"I would go further, and suggest that complex habitual sequences of motor output (e.g., drug seeking and drug consuming behavior in addicts) represent hypercomplex sequences of cortical-striatal- thalamic activation, triggered by sensory dependent amygdaloid-brainstem modulation of the monoaminergic systems during critical states.*

The power of ibogaine to break habitual patterns of addiction may reside in an induced SOC state that disrupts and functionally reorganizes this anygdaloid-brainstem system, in effect resetting the brain/mind." Again, we can suggest that CRP performs virtually the same function, without recourse to drug ingestion, thereby avoiding substitution of one chemically-induced experience for another.

Anderson thinks ibogaine works on many brain systems to *"drive firing dynamics into an SOC state with avalanches of phasic events similar to that existing during early development."* Clearly these same issues, experiences of ego-death, fragmentation and annihilation, and perinatal imagery, as well as all of the other classical transpersonal states of consciousness emerge spontaneously in CRP and are most often associated with spontaneous self-recovery on a variety of observational levels, indicating a fractal result, if not mechanism.

Biological systems are in a constant state of criticality and self-organization. Critical states in developing brains may lead to the enhancement of synaptic connections, sparing of axons, and synchronizing twitches that allow distant regions of the organism to link and coordinate gene expression and neural-motor development....Long patterns of bursting have statistical self-

similarity...they appear very similar to the bursting patterns of ion channels, neurons and phasic REM processes such as PGO waves.

Self-similar clusters in time result in unusual statistical properties, called Levy distributions. A unique property of these distributions is that they lead to what is called "*convolutional stability*." These are stable distributions self-similar over different sample sizes or time scales. Fractal clustering in the interconnected S-Net leads to knocking out the activity of some nodes and results in atypical fluctuations in 5-HT release in different brain regions.

These fractal fluctuations in S-Net activity may synergize with other neurotransmitter systems to bring new qualities to self-organized critical oneiric states, resulting perhaps in enhanced dopamine release in the amygdala and prefrontal areas. We might conjecture that an endogenous 5-HT reuptake inhibitor (SSRI) may prolong S-Net reorganization during therapy.

Fractal REM and Neural Plasticity

Most of the research on REM and fractal structure has been done only on sheep, but can be interpolated for our purpose. The REM-like sleep state is pervasive during fetal life, and development of the brain and behavior.

In 1996, Drs. Mandell and Anderson *"proposed that the correlated fractal bursting nature of REM, or Active sleep, as it is sometimes called in the fetus and newborn, provides an invariant Levy temporal framework in which cortical and subcortical networks can organize and consolidate changes."*

Findings show that phasic REM associated events, "at least during development, are not fundamentally independent random processes, as they are attributed in Allan Hobson's activation-synthesis model of REM sleep, but are rather fractal in time.

Anderson asserts that *"fractal Levy processes can be used to characterize the phasic events associated with fetal and adult REM sleep, such as eye movements, and may have great significance in understanding the relationship between REM sleep, neural plasticity, and ibogaine therapy."* Again, we would contend, similar statements can be made for CRP.

Anderson asks himself the same question which has piqued psychotherapist, Graywolf Swinney: *"Why do fetuses spend most of their time and energy in a state with strong similarities to adult REM sleep? Why is REM such an important component of our lives? In fact, REM sleep is almost as essential for life as water and food. Could the fractal properties of REM sleep in fetal animals provide a common thread between fetal and adult REM and insights into disorders of REM sleep?"*

CRP would suggest, it illuminates most all disorders, not only sleep disorders. For example, *"Disturbances of phasic REM processes are also a common thread in many disorders of sleep in infants, children and adults. As mentioned earlier, PTSD is linked to a fundamental disturbance of phasic REM sleep mechanisms resulting in recurrent stereotypical anxiety dreams as well as disturbed limbic system and brain stem-mediated functions such as abnormal startle responses. Chronic abuse of many drugs results in alterations of phasic REM sleep processes."*

Another avenue is that temporal lobe dysfunctions involving limbic structures such as the amygdala and hippocampus are frequently associated with sleep disturbances, and even sleep walking. Hemispheric asymmetries, resulting fro lateralized temporal lobe dysfunction and alterations of commissural development can be the aftermath of childhood stress or trauma.

Most theories of adult REMS function ignore its central role in fetal life. Most claim that adult and fetal REMS are too different to be considered relevant to adult behavior. But Anderson describes how alterations in the vertical and horizontal consolidation of self-similar bursting patterns of phasic sleep events can provide a conceptual bridge between the disorders of REM sleep in adults and in children. This conclusion underlies his hypothesis of bi-hemispheric reintegration.

Trauma or drug abuse history is strongly correlated with asymmetric hemispheric functioning. We have seen from Neurofeedback and Hemi-Synch research that this reintegration is fundamental to resetting the system back to healthful conditions, almost irrespective of the presenting disorder.

Amygdaloid stimulation evokes significantly increased PGO number, spike and burst density. Regional cerebral bloodflow in the human amygdala is positively correlated with REM sleep. The parabrachial region is also involved in alerting and in the generation of REM and PGO waves. Also cholinergic activation of the central nucleus produces long-term facilitation of REM.

The amygdala receives most of its serotonergic innervation from DRN which has a strong inhibitory influence upon amygdaloid neurons. Asymmetric activation of the amygdaloid-parabrachial pathways results in abnormal sleep architecture and pronounced changes in the patterns of phasic REM events. Eye movements in normal children on the whole do not become organized into bursts until 40 weeks gestational age; thereafter changes in the clustering of the bursts of eye movements (EM) are correlated with developmental age.

From 2-24 weeks postnatal, total REM decreases. Between 3 months - 5 years of age, a major organizational change occurs in the patterns of EMs, marked by the increasing tendency of bursts of EMs to cluster, with more and shorter EMs packed in bursts within bursts. Other research shows that maternally deprived neonatal rats showed a decrease in certain key fractal patterning.

Maternal deprivation, a model of early abuse, results in alterations of fractal measures of clustering, and these shifts persist into adult spontaneous activity. Early deprivation results in disruptions into adulthood, similar to symptoms of PTSD associated with child abuse.

Lasting changes in the amount or pattern of phasic activity associated with REM sleep appears to be one common thread linking PTSD and other sleep disorders in children and adults.

Functionally asymmetric cortical-amygdaloid-parabrachial pathways could also be a critical common factor in PTSD and other disorders, (Anderson).

Transformations in the quantities or characteristics of phasic REM processes may be a compensatory mechanism by which the brain attempts to reestablish horizontal and vertical consolidation through correlations inherent in the clustering process.

We allege that CRP treatment helps correct hemispheric asymmetry, phasic REM processes and the psychobiology of the amygdaloid complex. CRP Journeys are an experience of dream-like states, except that participants are awake and can respond. Images appear, especially after eyes are closed, often leading to rapid visual presentation of various images. All imagery arises organically in the dreamer; no metaphors are imported by the guide.

Often journeys lead to specific reviewing of traumatic events or circumstances from childhood and/or disorders. There are distortions of time perception, and the dream experience is perceived to take much less time than clock-time. Following re-entry, a period of intensive reevaluation of previous life experiences can take place.

The Pervasive Oscillatory Sound

Anderson speaks of the disturbing effect of lights and sounds, which might result from loss of normal global habituation due to RF destabilization, resulting in fear and/or rage. Again, trauma and drug abuse history is strongly associated with asymmetric hemispheric function. Temporal lobe structures such as the hippocampus and amygdala are particularly sensitive to the effects of child abuse and trauma.

Anderson conjectures the oscillatory sound could indicate rapid shifting or cycling of attentional resources between the left and right hemispheres, downshifting the normally constant 10 Hz rhythmicity of the olivocerebellar system. This oscillatory auditory effect may function as an auditory driver.

The downshift effect may indicate possible flooding of the left hemisphere by material from the uninhibited right which takes over primary conscious focus. This sets the stage, along with phasic fluctuations of the S-Net and uninhibited PGO, for the sudden onset of the SOC state and the waking dream period.

Anderson alludes to *"waking dreams as a healing journey through the fractal hyperspace of emotionally indexed childhood memories."* He asserts that *"the basolateral amygdala (BLA) is a critical neural substrate of the waking dream stage as fractal neural bursting in this subcortical cortex-like structure may represent access points in a fractal hyperspace of emotionally indexed memories."*

"The effects of early trauma on the development of the amygdala and other temporal lobe structures may interfere with its normal bilateral function during REM-sleep mediated consolidation of emotionally significant events. The recall of traumatic childhood experiences in

adults, due to the immaturity of limbic structures at the time of trauma, may require electrical stimulation or intensive PGO-like activity present during the oneiric state. Habitual disruption of normal sleep processes by stress associated with combat, bereavement, divorce, child abuse, neglect or chronic drug abuse interferes with the natural restorative function of phasic REM processes."

This exacerbates physiological and psychological addictions and rigidifies emotional traumas into PTSD and chronic hemispheric imbalance. CRP therapy may help to free these rigidities, restoring some degree of healthy hemispheric balance. The amygdala is the meeting place of emotions and the mind. We each have bilaterally interacting right and left amygdala which gives us our internal emotional experiences by processing and attaching affective response to the rich flow of information from all the five senses and modulating our perception of the autonomic centers in the brain.

Persinger insists that positive images (gods, angels, and light beings) emerge from the left lobe, while negative, daemonic imagery stems from the right. Echoing these attributions, Anderson adds that fear and anxiety are the most common feelings evoked frequently from the right amygdala. Direct electrical stimulation of the right BLA demonstrates that brain and mind meet to generate and bring to awareness the associated memories and emotions of a traumatic experience.

Conclusions

CRP and perhaps other process therapies, like ibogaine, may evoke the appropriate fluctuating milieu of neurotransmitters and neuromodulators to trigger a SOC state in the BLA, amygdaloid-brainstem pathways, and extrastriate areas activated during dreaming BLA cells have a unique morphology, pyramidal, or tetrahedrally omnidirectionally interwoven. This indicates the geometry of synergetics and an information-flow programmed for "the path of least resistance."

This connectivity is specialized for non-sequential interactions over multiple timescales, or broad-band synchronization. Distortions of time perception, Anderson thinks, may reflect the "rescaling in time" afforded by the fractal bursting of BLA pyramidal cells during this critical state. We can speculate that the role of common SOC states in the amygdalae, extrastriate cortex and brainstem form the emotional and visual substrates for CRPs "experiential" dream-like phenomena. After the abrupt end of the SOC dream-like state and rapid image experience, subjects are able to reflect on and integrate the experience.

The journeyer has experienced "the big picture" and has a unique perspective on his or her life. Experiential recall of trauma struggles helps bring resolution and getting in touch with soul, or a feeling of oneness with the universe, or other unitive expressions. CRP may function as a kind of facilitated "REM-rebound" process, making up for sleep loss since trauma or abuse first affected sleep architecture. After sojourners recover from their "journey to the land of the dead," they are reborn socially. But they return with the fractal perspective of the "long-view."

Although the brain generates long-range correlations, abuse, trauma and the cumulative stress of modern life can quickly destroy these correlations, so CRP is complimented with conventional forms of support, follow up, and reintegration. The traumatic abuse that may result in functionally abnormal hemispheric interactions precipitates in emotional instability and addictive behaviors.

CRP works through multiple neurotransmitter systems to create within amygdaloid-brainstem systems a self-organized critical oneiric state or state of plasticity, similar to states of plasticity existing during fetal development. This critical brain state may facilitate the consolidation of traumatic memories, reversal of abnormal hemispheric function and the dissolution of habitual motor patterns associated with addiction.

References

Anderson, C.M. (1998); "Ibogaine therapy in chemical dependency and posttraumatic stress disorder: a hypothesis involving the fractal nature of fetal REM sleep and interhemispheric reintegration," *MAPS*, Vol. 8 Number 1, Spring 1998, pp. 5-14.

Carhart-Harris et al, Robin L., Neural correlates of the psychedelic state as determined by fMRI studies with psilocybin, PNAS, 2012 [doi: 10.1073/pnas.1119598109].

Debru, C., 1990; *Neurophysiologie du reve (The Neurophysiology of Dreams)*, Paris, Hermann, Edition des Sciences et des Arts.

Edwards, Betty (1999); *[The New] Drawing on the Right Side of the Brain,* New York: Tarcher/Putnam.

Goutarel, N., Golnhoffer, N. and Siddens, R.(1993); "Pharmacodynamics and Therapeutic Applications of Iboga and Ibogaine; *Psychedelic Monographs & Essays,* Vol. 6; Boyton Beach, Florida: PM & E, 71-111.

Grof, Stanislav (1988) ; *The Adventure of Self-Discovery; Albany: SUNY Press.*

Jansen, Dr. Karl L. R. Jansen, MD, PhD, (2000), The Ketamine Model of the Near Death Experience: A Central Role for the NMDA Receptor, http://www.lycaeum.org/leda/docs/9264.shtml?ID=9264

Lakoff, George and Johnson, Mark (1980); *Metaphors We Live By* Lakoff, George (2000); *Philosophy in the Flesh*

Lakoff, George (1987); *Women, Fire, and Dangerous Things: What Categories Reveal About the Mind* . McKinney, Laurence O., (1994), Neurotheology: Virtual Religion in the 21st Century, Amer Inst for Mindfulness.

Miller, Iona and Swinney, Graywolf (2001); "CRP theta training: Theta reverie and co-consciousness in CRP; *Chaosophy 20001*, Wilderville: Asklepia Publications.

Miller, Iona and Swinney, Graywolf (2001); "The neuropsychology of CRP, Dreams, and

REM"; *Chaosophy 2001*, Wilderville: Asklepia Publications.

Murphy, Todd, describes phenomena of the cingulated gyrus and angular gyrus on Out-Of-Body experiences (http://www.innerworlds.50megs.com/obe.htm):

Nelson, J.P., R. W. McCarley, and J. A. Hobson, (1983), REM sleep burst neurons, PGO waves, and eye movement information, http://jn.physiology.org/content/50/4/784

Nichols, David E. (1998), "The Heffter Review of Psychedelic Research, Volume 1, 1998 - 5. The Medicinal Chemistry of Phenethylamine Psychedelics.

Othmer, Siegfried, Othmer, S., Kaiser, David; "EEG biodfeedback: A generalized approach to Neuroregulation"; in*Applied Neurophysiology & Brain Biofeedback,* Ed. Kall, Kamiya, and Schwartz.

Penrose, PhD, Roger, and Stuart Hameroff, MD, (2011), Consciousness in the Universe: Neuroscience, Quantum Space-Time Geometry and Orch OR Theory, Journal of Cosmology, 2011, Vol. 14. http://www.quantum-mind.org/cosmology160.html

Sabom, M.B. (1982); *Recollections of Death*, New York:Harper & Row. Schiffer.

Strassman, Richard (1991); "The Pineal Gland: Current Evidence for its Role in Consciousness"; *PM & E,* Vol. 5, 167-205.

Swinney, Graywolf and Miller, Iona (1992); *Dreamhealing,* Wilderville: Asklepia Pub. Swinney,

Swinney, Graywolf (1999);*Holographic Healing;* Wilderville: Asklepia Pub.

Ramachandran, (2005)
http://www.eurekalert.org/pub_releases/2005-05/uoc--gmu052005.php

Rossi, Ernest (2000); "Sleep, dream, hypnosis and healing: behavioral state-related gene expression and psychotherapy," in *Sleep and Hypnosis* 1:3, 1999, pp. 141-157.

Rousseau, D. (2012) The Implications of Near-Death Experiences for Research into the Survival of Consciousness, Journal of Scientific Exploration, Vol. 26, No. 1, pp. 43-80.

Rousseau, D. (2011) Minds, Souls and Nature: A Systems-Philosophical Analysis of the Mind-Body Relationship in the Light of Near-Death Experiences . (PhD Thesis, University of Wales, Trinity Saint David, School of Theology, Religious Studies and Islamic Studies).

Rousseau, D. (2011) Near-Death Experiences and the Mind-Body Relationship: A Systems-Theoretical Perspective, Journal of Near-Death Studies, 29, pp. 399-435.

Rousseau, D. (2011) Understanding Spiritual Awareness in Terms of Anomalous Information Access, The Open Information Science Journal - Special Issue: Information and Spirituality, 3, pp. 40-53.

Szechtman, Henry, Erik Woody, Kenneth S. Bowers, and Claude Nahmias§, Where the imaginal appears real: A positron emission tomography study of auditory hallucinations
http://www.pnas.org/content/95/4/1956.full

Tart, Charles; (2001) *States of Consciousness*, Universe (January 9, 2001).

Tart, Charles, (1990), *Altered States of Consciousness*, Harper; 3rd Revised edition (1990).

Article

Persinger Group's Recent Experiments, Spin Network and TGD

Matti Pitkänen [1]

Abstract

Michael Persinger's group reports three very interesting experimental findings related to EEG, magnetic fields, photon emissions from brain, and macroscopic quantum coherence. The findings provide for the proposal of Hu and Wu that nerve pulse activity could induce spin flips of spin networks assignable to cell membrane. In this article I analyze the experiments from TGD point of view. It turns out that the experiments provide support for several TGD inspired ideas about living matter. Magnetic flux quanta as generators of macroscopic quantum entanglement, dark matter as a hierarchy of macroscopic quantum phases with large effective Planck constant, DNA-cell membrane system as a topological quantum computer with nucleotides and lipids connected by magnetic flux tubes with ends assignable to phosphate containing molecules, and the proposal that "dark" nuclei consisting of dark proton strings could provide a representation of the genetic code. The proposal of Hu and Wu translates to the assumption that lipids of the two layers of the cell membrane are accompanied by dark protons which arrange themselves to dark protonic strings defining a dark analog of DNA double strand.

1 Introduction

Michael Persinger's group reports [5, 6, 7] three very interesting experimental discoveries relating to EEG, magnetic fields, photon emissions from brain, and macroscopic quantum coherence.

In the first article [5] entitled *Congruence of Energies for Cerebral Photon Emissions, Quantitative EEG Activities and* $\sim 5\ nT$ *Changes in the Proximal Geomagnetic Field Support Spin-based Hypothesis of Consciousness* correlations between cerebral photons emissions, EEG, and changes of the proximal geomagnetic field are reported. The findings provide support for the proposal of Hu and Wu [8] that nerve pulse activity could induce spin flips of spin networks assignable to cell membrane motivated by the observation that the magnetic spin-spin interaction between protons at a distance of 10 m (cell membrane thickness) corresponds to energies for which frequency is in EEG range.

In the second article [6] entitled *Demonstration of Entanglement of Pure Photon Emissions at Two Locations That Share Specific Configurations of Magnetic Fields: Implications for Translocation of Consciousness* the group reports an excess correlation between "pure" photon emissions at two locations separated by few meters that share specific correlations of frequency modulated magnetic fields. The photon emissions were from LEDs in the experiment consider. In an earlier similar experiment, which is also discussed, they were from chemical reactions occurring in solutions contained by cell cultures.

In the third article [7] entitled *Experimental Demonstration of Potential Entanglement of Brain Activity over 300 Km for Pairs of Subjects Sharing the Same Circular Rotating, Angular Accelerating Magnetic Fields: Verification by s_LORETA, QEEG Measurements* an excess correlation of brain activity of subject persons separated by 300 km and sharing the same circular rotating, angular accelerating magnetic fields is reported.

It turns out that the experiments provide support for several TGD inspired ideas about living matter. Magnetic flux quanta as generators of macroscopic quantum entanglement, dark matter as a hierarchy of macroscopic quantum phases with large effective Planck constant, DNA-cell membrane system as a topological quantum computer with nucleotides and lipids connected by magnetic flux tubes with ends assignable to phosphate containing molecules, and the proposal that "dark" nuclei consisting of dark

[1] Correspondence: Matti Pitkänen http://tgdtheory.com/. Address: Köydenpunojankatu 2 D 11 10940, Hanko, Finland. Email: matpitka@luukku.com.

proton strings could provide a representation of the genetic code. The proposal of Hu and Wu [8] translates to the assumption that lipids of the two layers of the cell membrane are accompanied by dark protons which arrange themselves to dark protonic strings defining representation for DNA sequences.

In the sequel I briefly explain my own interpretation of these experiments and their outcomes from TGD point of view and show that a nice interpretation of the findings emerges. Before going to this it is however appropriate to summarize briefly those aspects of TGD based view about living matter, which are relevant for the interpretation of the experiments.

1.1 Key aspects of the TGD inspired vision about living matter

The key ingredients of TGD inspired vision about living matter needed in the sequel are following.

1. The notion of many-sheeted space-time is the first new element [19, 20]. Space-times are 4-D surfaces in 8-D space-time $M^4 \times CP_2$ so that one has what might be called sub-manifold gravity. Any physical system corresponds to a space-time sheet characterizing its shape and size. The outer boundaries of macroscopic objects correspond to causal boundaries at which the signature of the induced metric of the space-time surface changes. Therefore space-time surfaces are topologically non-trivial in all scales and we directly perceive it. Space-time surfaces form a fractal hierarchy in the sense that subsystems of system correspond to space-time sheets topologically condensed at it via the formation of wormhole contacts which are regions of space-time with an Euclidian signature of the induced metric.

   Also the notion of classical field is topologized. Various classical fields are subject to what might be called topological field quantization. For instance, radiation fields decompose to topological light rays and magnetic field to magnetic flux quanta (flux tubes and flux sheets). Topological field quantization is of special importance in living matter and leads to the notion of field body and magnetic body as additional structural and functional parts of a living system.

2. p-Adic physics [26] defines a further basic element. p-Adic number fields are proposed to serve as correlates for cognition in the sense that one can speak about p-adic space-time sheets as correlates for cognition and for intentions [23, 28] The quantum jump transforming p-adic space-time sheet to a real one corresponds to a transformation of intention to action. The generation of though in turn corresponds to an opposite of this transition. Zero energy ontology makes this picture internally consistent and no breaking of conservation laws is implied.

   p-Adic length scale hypothesis [24] states that p-adic primes near powers of 2 are of special physcal significan and Mersenne primes $M_n = 2^n - 1$ especially so. A possible explanation for the importance of these primes is that evolution corresponds to a gradually increasing complexity. These primes are simple in the sense that all digits in their pinary expansion are '1':s expect possible some for the first few ones) are especially interesting physically because they should have emerged first. Mersenne primes have only '1':s in their pinary expansion so that they are the simplest possible primes and indeed seem to correspond to fundamental physical scales. This leads to quite powerful predictions in particle physics context.

   In the scales of living matter a number theoretical miracle occurs: in the length scale range from 10 nm (cell membrane thickness to 2.5 μm (size scale of cell nucleus) as many as four Gaussian Mersenne primes $M_{G,n} = (1+i)^n - 1$ occur and correspond to p-adic primes near p^k, $k = 151, 157, 163, 167$.

3. The hierarchy of *effective* Planck constants [18] coming as integer multiples $\hbar_{eff} = n\hbar$ of the ordinary Planck constant was partially motivated by the findings of Blackman [3] and others related to the unexpected effects of ELF em fields on vertebrate brain. These effects look quantal but this should not be possible since the cyclotron energies in the magnetic field $.2 \times 10^{-4}$ T (2/5 times the nominal value of the Earth's magnetic field $B_E = .5 \times 10^{-4}$ T) are 10 orders of magnitude below the thermal threshold.

This led to the hypothesis about the value spectrum of Planck constants. The phases of ordinary matter with non-standard value of effective Planck constant are identified as dark matter. Later two different - possibly equivalent - reductions of the hierarchy to that for *effective* values of $\hbar$ have emerged in TGD framework [28].

One of the most speculative ideas related to the dark matter hierarchy is based on the observation that a simple model for dark proton implies that the states of dark proton are in 1-1 correspondence with DNA, RNA, tRNA, and amino-acids, and that there is a simple rule reproducing vertebrate genetic code [22, 21]. Dark nuclei defined by sequences of dark protons would define the analogs of DNA sequences so that genetic code would not be a outcome of random bio-chemical selection but a basic element of particle physics, and biological systems would only define a secondary representation of the fundamental genetic code. This proposal has far reaching implications. Surprisingly, the findings of the first article [5] supporting the hypothesis of Hu and Wu [8] about proton spin networks combined with the dark DNA hypothesis lead to a concrete model for the proton spin networks as paired dark DNA sequences assignable to the two lipid layers of the cell membrane.

4. Magnetic flux tubes carrying dark matter take central role in TGD inspired quantum biology. The knotting and braiding of the flux tubes makes possible topological quantum computation and leads to the hypothesis that DNA and cell membrane connected by flux tubes form a topological quantum computer [17]. Flux tubes can connect sub-systems of living organisms or even different organisms to form coherent structural and functional units. Indeed, the large value of $\hbar_{eff}$ makes possible macroscopic quantum coherence. In particular, biomolecules can be connected by flux tubes to coherent structures. The reconnection of flux tubes plays a key role in the proposed model bio-chemical reactions and bio-catalysis. Inportant are also the phase transitions changing the value of Planck constant inducing in turn a change of the length of the flux tube identified as a quantal length scale depending of $\hbar_{eff}$. These phase transitions could be responsible for the phase transitions changing dramatically the density of matter in cellular interior (say sol-gel transition).

   Cyclotron Bose-Einstein condensates at magnetic flux tubes are proposed to be a characteristic of living systems [12]. Cyclotron frequencies are classical (no dependence on Planck constant) but cyclotron energies scale like $\hbar_{eff}$ so that for a large enough value of the effective Planck constant cyclotron energies of dark photons are above thermal threshold, and can induce macroscopic quantum coherence. Dark photons decay to bunches of ordinary photons and an attractive hypothesis is that bio-photons result as decay products of dark photons.

   The notion of magnetic body emerges naturally. Any physical system is accompanied by magnetic fields which in TGD Universe defines separate entity, which can be called magnetic body. Magnetic body is identified as an intentional agent using biological body as sensory receptor and motor instrument. Magnetic body has an onion like structure corresponding to the hierarchy of space-time sheets defining physical system, say biological body. The size of the magnetic body is much larger than that of biological body. 10 Hz frequency corresponds to a layer with size large than the size scale of Earth.

5. Zero energy ontology (ZEO) [11] is a further basic element. In zero energy ontology physical states are zero energy states consisting of pairs of positive and negative energy states having opposite net quantum numbers and being localized to the opposite light-like boundaries of $CD \times CP_2$, where CD is causal diamond identified as an intersection of of future and past directed light-cones and defining a structure analogous to double pyramid (a convenient shorthand for $CD \times CP_2$ is simply CD).

   The interpretation of zero energy states is as counterparts of pairs of initial and final states of physical events in positive energy ontology. CDs form a fractal hierarchy with CDs within CDs. The size scales of CDs come as integer multiples of CP_2 size scale about 10^4 Planck lengths. One can interpret CD as an imbedding space correlate for a "spot light of consciousness" in the sense that the conscious experience of self associated with given CD is about region defined by CD. Space-time sheets within CD serve as correlates for selves at space-time level.

Also elementary particles are expected to be accompanied by CDs, and one especially important prediction is that the time scale of the CD associated with electron is .1 seconds, which corresponds to the fundamental 10 Hz bio-rhythm. All elementary particles correspond to macroscopic time scales and u and d quarks would correspond to time scales between 1 ms and .1 seconds.

1.2 Cell membrane as super-conductor and a model for EEG

The proposal is that cell membrane is accompanied by super-conducting dark magnetic flux tubes [25]. Cooper pairs of electrons, protons, and biologically important fermionic ions would be the carriers of supra currents besides bosonic ions such as Ca^{++} and Mg^{++}. Note that the new exotic nuclear physics suggested by TGD allows to imagine that fermionic nuclei could appear as bosonic variants with essentially same chemical properties [22].

Josephson currents through cell membrane have frequency $f = eV/\hbar_{eff}$ so that in this case the energy $E = eV$ identifiable as the energy of electron or proton gained in traversing the cell membrane is classical quantity whereas Josephson frequency is quantal [25]. Situation is the opposite of this for cyclotron frequencies and energies. Obviously, large values of $\hbar_{eff}$ correspond to low Josephson frequencies. Soliton sequencies associated with the Sine-Gordon equation governing the dynamics for small variations of membrane potential would represent ground states of axonal membranes mathematically analogous to a sequence of mathematical penduli rotating in phase. Nerve pulse generation would mean a perturbation in which one pendulum is kicked [25].

There are two alternative models for the cell membrane as a Josephson junction [12].

1. For the conservative option [12] the cell membrane is far-from vacuum extremal and various charged particles experience only the electromagnetic field. The energy scale of excitations is determined by the electric voltage and is given by $E = eV$. Nerve pulse generation would be associated with this kind of membranes. Josephson radiation with harmonics of $f = eV/\hbar_{eff}$ is one signature of super-conductivity.

   One ends up also to an explanation of EEG in this framework [16]. The function of EEG would be communication of sensory data from cell membrane to the magnetic body and control of biological body via flux sheets traversing through DNA, where genetic expression is activated by the control signals. EEG frequencies are linear combinations of harmonics of Josephson frequencies and of the increments of cyclotron frequencies. Cyclotron transitions can be also accompanied by a spin flip. This model allows to identify EEG bands. The hierarchy of Planck constants suggest a generalization of EEG and its variants (say EKG) to a fractal hierarchy obtained by scaling EEG. For large enough values of $\hbar$ cyclotron contributions to EEG energies would correspond to energies above thermal threshold as also Josephson frequency ($E = eV_{thr}$, where V_{thr} is the value of resting potential at which nerve pulse is generated, is just at the threshold). This would make possible the correlation of EEG with the brain state and also quantum biocontrol by using photons with EEG frequencies.

2. For the non-conservative option [15] cell membrane is near- to vacuum extremal. The classical Z^0 fields predicted by TGD dominate over em fields, and the voltage must be replaced by a combination of Z^0 and em voltages. By assuming that the Weinberg angle is considerably smaller in this phase than in the standard phase the energies gained by various ions correspond to visible photons. This hypothesis allows to understand the frequencies for which photoreceptors - which do not directly generate nerve pulses - are most sensitive. Near vacuum extremal property obviously implies high sensitivity to perturbations making sensory receptor optimal.

An interesting possibility is that the far-from *resp.* near-to vacuum extremal options are realized for the neurons of left *resp.* right hemisphere. This option finds support from the observation of Persinger et al [5] that visible photon emissions are mostly from the right hemisphere. Another possibility is that glial cells as cells which do not generate nerve pulses correspond to near-to vacuum extremals. The identifications do not exclude each other.

1.3 Learning to apply the notion of induced field

The geometrization of classical gauge fields and gravitational fields relying on the induction of spinor connection of CP_2 and $M^4 \times CP_2$ metric to the space-time surface is one of the key ideas of TGD and it is useful to get more concrete understanding of the induced fields since this notion will be applied in the sequel.

1.3.1 The basic objection and its resolution

The basic objection against the induced fields is that they reduce the dynamics to that of only 4 field like variables since the 8 imbedding space coordinates take the role of field variables and 4 of them are eliminated by general coordinate invariance as field variables. Besides this preferred extremals of Kähler action represent space-time surfaces carrying very restricted kind of patterns of induced gauge fields analogous to Bohr orbits.

Many-sheeted space-time however saves the situation. Each system creates its own field body represented in terms of topological field quanta. If these field bodies have common M^4 projection, test particle topologically condense to each of these field body (touches each of them), and the effects of these fields sum up although fields do not interfere as they would do in ordinary field theory.

1.3.2 How could one generate dark photons with large $\hbar$?

The observation which led to the proposal of the effective hierarchy of Planck constants, was that microwaves with frequency of f_h modulated by ELF frequency f_l induce in vertebrate brain effects which could be understood in terms of cyclotron frequencies assignable to quantal cyclotron transitions in and endogenous magnetic field for which cyclotron frequency was equal to ELF frequency: $f_c = f_l$. These effects are possible only if the cyclotron energy is above thermal energy, and this led to the proposal about the hierarchy of Planck constants. '

The key question is how the modulation by ELF frequency could generate dark photons with large $\hbar_{eff}$. A possible answer to this question comes from another question. Topological field quantization forces to ask what amplitude modulation of fields means.

The simplest modulation corresponds to a multiplication of rapidly oscillating field with a slowly varying oscillating amplitude so that amplitudes with frequencies $f_h \pm f_l$ result ('h' and 'l' refer to "high" and "low"). The natural thing to do is to develop the two amplitudes with frequencies $f_h \pm f_l$ in Fourier series in time interval $T = 1/f_l$. All harmonics of f_l appear and coefficients of the expansion are proportional to $1/(f_h - (n \pm 1)f_l)$. Maximal amplitudes correspond to $f_h \simeq (n \pm 1)f_l$. This suggests that when this almost resonance condition is satisfied the generation of dark photons with frequency f_l and energy $\hbar_{eff} f_l$, with $\hbar_{eff} \simeq f_h/f_l$, can take place with a considerable rate. If this argument is correct, one could generate dark photons with given $\hbar_{eff}$ by using modulation satisfying the condition $f_h/f_l = \hbar_{eff}$.

In the case of ELF em fields interacting with brain this is not enough since microwave photons have energies below the thermal threshold E_{th}. Bio-system however contains photons with energy above thermal threshold - say bio-photons with frequencies f in visible range or infrared Josephson photons generated by cell membrane Josephson currents - the fields associated with MEs ("massless extremals", topological light rays) accompanying these many-photon states can be modulated by the ELF modulated microwaves. Since one can say that a modulation of modulation is modulation, the outcome is modulation (f, f_{ELF}) producing dark photons with $\hbar_{eff} \simeq f/f_{ELF}$ with energies about E_{th}. This mechanism would explain the "scaling law of homepathy"' [21] stating that fields with low frequencies f_l are somehow transformed to fields with high frequencies f_h and vice versa. The proposal has been that large $\hbar_{eff}$

photons with $\hbar_{eff} \simeq f_h/f_l$ decay to ordinary photons or vice versa. This transformation has quite concrete description $\hbar_{eff} = n$ photons corresponds correspond to n-furcations of space-time surface made possible by the non-determinism of Kähler action. All n-sheets of the n-furcation would be present and each of them would carry photon with frequency f_l and total energy would be $\hbar_{eff} f_l = f_h$.

1.3.3 How to describe time varying magnetic fields?

The topological flux quantization for static magnetic fields is easy to understand. The description of time varying magnetic fields in terms of flux quanta is however a non-trivial exercise in thinking in terms of topological field quanta.

Flux quantization implies that the magnetic dipole field decomposes into closed flux tubes with a straight part inside dipole and a portion outside the dipole carrying return flux in roughly opposite direction also arranged to flux tubes.

The basic assumption is that the flux tube structure of dipole field is not lost but is only re-arranged as the dipole field oscillates. As the dipole strength decreases the flux tubes along field lines outside the dipole contract so that eventually the closed flux tubes of dipole field degenerate to those of wormhole magnetic fields [27] restricted inside the dipole and consisting of parallel flux tube space-time sheets with same M^4 projection and carrying opposite magnetic field strength and having distance of order CP_2 length along CP_2 direction. A charged particle topologically condensing at both sheets experiences the sum of the magnetic fields, which vanishes. As the sign of dipole changes, the flux tubes in the interior of dipole begin to move to the exterior of the dipole. In operational sense this dynamics is approximated well by Maxwell's theory or vice versa.

How the electric electric fields associated with the time varying magnetic field predicted by Faraday law are represented? These fields are rotational with flux lines rotating around the magnetic field. In Maxwell's theory one would have single vortex like structure. In TGD this vortex like structure decomposes into smaller vortices assignable to individual flux tubes just like the rotational flow of superfluid decomposes into smaller vortices satisfying quantization condition analogous to the quantization of the magnetic flux.

Also the geometro-dynamics for the flux quanta of electric field is possible and in this case magnetic fields induced by time dependent electric fields are assignable to flux quanta. Cell membrane is a good example of this kind of situation. Quite generally, the geometro-dynamics of topological field quanta together with the possibility to have varying overlapping M^4 projections allows to reproduce the smooth dynamics of Maxwell fields.

2 First article

The first article has the title *Congruence of Energies for Cerebral Photon Emissions, Quantitative EEG Activities and* $\sim 5\ nT$ *Changes in the Proximal Geomagnetic Field Support Spin-based Hypothesis of Consciousness*, which already summarizes the findings.

2.1 Findings

In the article [5] Persinger's group reports simultaneous changes in photon emissions, EEG activity, and alternations of geomagnetic field when a person sitting in dark is imagining white light or not. The abstract of the article is following.

The hypothesis by Hu & Wu that networks of nuclear spins in neural membranes could be modulated by action potentials was explored by measurements of the quantitative changes in photon emissions, electroencephalographic activity, and alterations in the proximal geomagnetic field during successive periods when a subject sitting in the dark imagined white light or did not. During brief periods of imagining white light the power density of photon emissions from the right hemisphere was about $10^{-11}\ Wm^{-2}$ *that was congruent with magnetic energy within the volume associated with a diminishment of* $\sim$ *7 nT as predicted by the dipole-dipole coupling relation across the neuronal cell membrane. Spectral analyses showed maxima in power from electroencephalographic activity within the parahippocampal region and photon emissions from the right hemisphere with shared phase modulations equivalent to about 20 ms. Beat frequencies (6 Hz) between peak power in photon (17 Hz) and brain (11 Hz) amplitude fluctuations during imagining light were equivalent to energy differences within the visible wavelength that were identical to the intrinsic 8 Hz rhythmic variations of neurons within the parahippocampal gyrus. Several quantitative solutions strongly suggested that spin energies can accommodate the interactions between protons, electrons, and photons and the action potentials associated with intention, consciousness, and entanglement.*

The authors interpret the results in terms of entanglement identified as enhanced correlations. Entanglement in this sense does not correspond to quantum entanglement. To my opinion (quantum) coherence would be a more standard manner to interpret the findings. Quantum coherence of course makes possible also quantum entanglement.

Spin flips, whose importance for consciousness has been emphasized by Hu and Wu [9]. The spin flips would occur between spin triplet and singlet states of pairs of protons belonging to the spin network. The basic finding is that the energy changes are accompanied by changes in EEG power.

Note that spin flips are possible also for cyclotron states proposed to be important for consciousness in TGD approach. In the case of electron the change of the energy in spin flip is in excellent approximation the same as in the transition $n \rightarrow n \pm 1$ of cyclotron state characterized by integer n (radial wave functions of electron in constant magnetic field correspond to those of harmonic oscillator). For ions the Lande factor g characterizes the effective nuclear angular momentum and appears in the spin flip energy and also now the frequencies involved are in EEG range.

The correlation of photon emissions with imagination of white light supports the hypothesis that EEG photons are responsible for communications to and control of biological body by magnetic body.

2.2 TGD inspired interpretation of the findings

What has been observed is correlation between EEG, emission of visible photons, and weakening of Earth's magnetic field with the change of magnetic energy equal to the energy of radiated photons. There is also evidence that spin flip transitions for protons are involved.

2.2.1 What is the origin of the visible photons?

The basic question concerns the origin of the visible photons.

1. An attractive general hypothesis is that the visible photons result in the transformation of dark EEG photons to ordinary visible photons. In TGD based model EEG (and its predicted fractal variants) correspond to dark photons with large *effective* value of $\hbar$ - call it $\hbar_{eff}$ - and energy $E = h_{eff} f$ in infrared or visible range and perhaps even in UV. Also bio-photons would result from these large $\hbar$ "dark" photons as they decay to bunches of ordinary photons. The wavelengths of dark photons with given energy are scaled by $\hbar_{eff}/\hbar$ predicted to be integer. The transformation of EEG photons to ordinary visible photons could explain the correlation between EEG and visible photon emission reported by Persinger's group. This kind of process would generate also biophotons.

2. The mechanism providing energy for dark photons (in particular EEG photons) would provide it also for the visible photons. According to the authors, the energy would come from the Earth's magnetic field which I as inhabitant of many-sheeted space-time take liberty to translate to "measured magnetic field". What is interesting that magnetic body would serve as a provided of metabolic energy. It is interesting to notice that in TGD based cosmology matter is created from the dark energy identified as Kähler magnetic energy assignable to magnetic flux tubes.

3. Authors conclude that the energy liberated per action potential is $E = eV_{rest}$. In TGD framework it could correspond to either a photon of Josephson radiation or the energy liberate when electron traverses the cell membrane. What is troublesome that this energy corresponds to IR photon just above thermal threshold rather than visible photon. The non-conservative model for the cell membrane mentioned above (applying to photo-receptor cells at least) could explain why visible photons rather than infrared photons with energy $E = eV_{rest}$ correspond to photons of the Josephson radiation.

4. The model based on the observation of Hu and Wu [8] suggesting that action potentials affect a spin network of protons (possibly at opposite ends of lipid of two lipid layers making cell membrane) looks like a totally different explanation from what would come first in mind in TGD framework. Could the spin network proposal of Hu and Wu be integrated to the picture of living matter provided by TGD? This is the question to be considered next.

2.2.2 The spin network hypothesis of Hu and Wu from TGD view point

The hypothesis of Hu and Wu [8] states that nuclear spin networks of nuclei associated with the cell membrane are relevant for consciousness in the sense that action potential induces modulations of the coupling parameters describing the magnetic interaction between neighboring spins of the spin network.

1. A direct calculation using the value of proton magnetic moment gives that the magnetic field created by proton at distance defined by cell membrane thickness of 10 nm is 3 nT. There are also other factors involved, and the estimate of Hu is that the field is about 5 nT.

2. The crucial observation is that the classical spin-spin interaction energy for two protons at distance $d = 10$ nm defined by cell membrane thickness and given by $E_{s-s} = -\mu \cdot B$, where B is the dipole field created by proton, corresponds to a frequency of the order 10^{-14} eV and thus is in EEG range. This can be seen by a direct calculation by assuming that proton creates a dipole field with Lande factor of proton.

   The frequencies assignable to the energies of neighboring interacting proton spins at distance d are in EEG range also when the effects of the environment are taken into account. For instance, the Hamiltonian for a rotationally symmetric nearest neighbor spin-spin interaction characterized in terms of so called J-factor, predicts in the case of protons frequency differences ΔE between singlet and triplet states varying in the range 5-25 Hz.

   For heavier nuclei these interaction energies scale down like $1/A^2$, A the mass number, so that a naive conclusion would be that the frequencies tend to be below 5 Hz scale. Proton would therefore be in a completely unique position. That EEG frequencies result in case of proton suggest that cell membrane thickness is not 10 nm by a pure accident (not that p-adic length scale hypothesis fixes assigns it to the p-adic length scale $L(k = 151)$, where $k = 151$ characterize Gaussian Mersenne prime.

The fact that the frequencies for energy differences of singlets and triplets are in EEG range is highly relevant also from TGD point view since this energy range makes it possible for EEG frequencies to induce spin flips.

1. In TGD framework fermionic spin and fermion numbers in various modes of second quantized induced spinor field (1 or 0) are predicted to serve as correlates for Boolean cognition [14] so that there are good reasons to expect that also spin flips are important. One might even think that protonic and even nuclear spins could be utilized to build Boolean representations.

2. The basic objection against the proposal of Hu and Wu is same as that against the findings of Blackman and others: quantum coherence is not possible since the energy differences corresponding

to (say) frequency of 5 Hz is about 12 orders of magnitude below thermal threshold. Trom the basic relation $E = \hbar_{eff} f$ it is clear that the objection can be circumvented for large values of effective Planck constant, which can take raise the energies involved to those of IR or perhaps even visible photons.

3. Authors conclude that the energy emitted per single action potential is $E = eV_{rest}$ which corresponds to IR photon just at the thermal threshold. It is however visible photons which are emitted. Why not photons with the Josephson energy $E = eV_{rest}$ just at the thermal threshold?

   If the photons would result when electron or proton traverses cell membrane and liberates potential energy as a photon or if the emitted IR photon could be interpreted as a photon of Josephson radiation this would be the case. TGD allows also to imagine that the cell membranes in question correspond to the non-conservative option for the model of cell membrane as Josephson junction for which V_{rest} contains Z^0 potential as a dominating contribution and gives rise to Josephson photons with energies in visible range.

   If one takes the proposal of Hu and Wu seriously, the visible photons would have different origin, and one must perhaps give up the assumption that the estimate of authors forces the identification of basic energy quantum emitted in the process considered as $E = eV_{rest}$.

Authors state that the energy associated with visible photon emission should be equivalent to the energy emitted in the emission of photons. What can one conclude from this?

1. An attractive possibility would be "dark" spin network formed by spin-coupled protons, whose members are associated with the lipids of the two lipid layers with lipids. The number of the lipids per cell membrane would be roughly $N_l = r^2/d^2$, with lipids thickness estimated to be $d \sim .1$ nm. For $r \sim 10^{-4}$ m corresponding to a relatively large neuron this would give $N_l = 10^{14}$.

   This number would give also the maximum number of spin pairs participating in phase transition and an estimate for the value of $\hbar_{eff}$ from $N_l \Delta E_{s-s} = E_{ph}$ as

$$N_l = \frac{E_{ph}}{E_{s-s}} = \frac{f_{ph}}{f_{s-s}} \quad .$$

   Suppose that all dipoles make a simultaneous spin flip with energy change $\Delta E = hf$, $f_{s-s} = 5$ Hz generating an energy of $E_{ph} = 1eV$ corresponding to a frequency of 2.4×10^{14} Hz. This requires $N_l \sim .5 \times 10^{14}$. It is encouraging that the rough estimates are consistent with each other.

2. That all protonic spin pairs make a simultaneous spin flip between singlet and triplet states of neighboring pairs looks like a phase transition. This suggests strongly macroscopic quantum coherence. What looks extremely strange that single visible photon should be emitted in the process since the entire magnetized region would behave like single spin! In standard physics this is not possible. TGD however leads to a possible realization of this kind of process as a mechanism of psychokinesis [29].

   The hierarchy of effective Planck constants could resolve the paradox. If one has $\hbar_{eff}/\hbar \simeq E_{ph}/\Delta E \simeq .5 \times 10^{14}$, the emitted photon would be large $\hbar$ dark photon with frequency 5 Hz and the energy of visible photon and geometrically would corresponds to a n-furcation of space-time with $n = \hbar_{eff}/\hbar$ sheets each carrying single 5 Hz photon. Each dipole pair emits ELF photon but they combine to single dark ELF photon with the energy of single photon.

It seems that it is not natural to assign the photon emission to cyclotron transitions ionic cyclotron B-E condensates or to the transitions associated with the cell membrane Josephson junctions. Also the model based on the observation of Hu and Wu is very attractive. This does not add a completely new element to TGD. One can find a nice connection with one of the TGD inspired basic ideas about genetic code, namely the dark realization of genetic code as sequences of dark protons.

1. For about 7 years ago I constructed a model for dark nuclei identifying the as strings of dark nucleons [22, 21]. The model of dark nucleon yielded a compete surprise: the states of the nucleon were in 1-1 correspondence with DNA, RNA, tRNA, and amino-acids and vertebrate genetic code could be understood in simple manner. This led to the vision that dark proton sequences allow a virtual world realization of genetic code making possible a kind of R&D department developing and testing various genetic alternatives. The genetic discoveries are however useful only if they can be used. This requires a generalization of transcription process allowing to transcribe DNA and RNA and perhaps even tRNA, and amino-acids to their dark counterparts and vice versa. This requires that dark nucleon sequences have same size scale as ordinary DNA, RNA, and amino-acids and that they could accompany the biomolecules.

   This fixes the size scale of dark proton to be of the order of the volume defined by the length L corresponding to single nucleotide in nucleotide sequence. The value of Planck constant would be of the order $\hbar_{eff}/\hbar \sim L/r_p \simeq 2.3 \times 10^5$, $r_p = \hbar/m_p \simeq 1.3 \times 10^{-15}$ m and$L \simeq .3$ nm.

2. At the same time I also constructed a model of DNA and cell membrane acting as a topological quantum computer [17]. DNA nucleotides would be connected to lipids of the inner lipid layer of the cell membrane by magnetic flux tubes, whose braiding would define the topological quantum computer programs. The braids would continue from the outer lipid layers to the membranes of other cells and in this manner bind the cells to a kind of network. The strands could have at their ends molecules containing phosphates to make possible transfer of metabolic energy to the system.

3. Dark protons could be generated in the ionization of OH group to OH^- as proton drops to dark space-time sheet and possibly becomes a part of dark proton sequences.

   (a) The basic process would be formation of dark water in this manner and the rich spectrum of anomalies of water could be understood in terms of temperature dependence fraction of dark protons [15].

   (b) OH groups are also associated with the hydrophilic ends of lipids such as fatty acids, glycerolipids, and phospholipids, which are the basic structural element of cell membranes. In phospholipids OH is associated with phosphate. In the DNA strand the phosphates contain O^- identifiable as OH^- resulting when proton of H drops to dark space-time sheet and possibly becomes part of dark proton sequence.

   (c) Also carbohydrates, in particular sugars, which are basic building brick of metabolism and defined the sugar backbone of DNA and RNA, contain a large number of OH groups. The model of DNA as topological quantum computer led to a proposal that magnetic flux tubes have OH or OH^- groups as their ends. These observations would allow magnetic flux tubes have dark protons at either or both ends. According to the earlier proposal [17] magnetic flux tubes to have OH and $O=$ at their ends. Earlier picture need not to be modified if the cell membrane carries dark double DNA strand connected to the ordinary DNA double strand inside nucleus. Similar connections would be natural also between DNA and amino-acids and their dark counterparts possibly associated with the cell membrane and reconnection of the color magnetic flux tubes could allow to build and manipulate these connections.

4. This would predict that single DNA codon, which corresponds to a length of .33 nm along DNA strand is connected to single lipid by magnetic flux tube or three color magnetic flux tubes to corresponding proton consisting of 3 quarks. This seems to be consistent with the width of single lipid in lipid bilayer if one takes seriously the illustration of the Wikipedia article. Note that in the earlier model single nucleotide was assumed to be connected by a magnetic flux tube to *single lipid*.

5. A further natural working hypothesis is that the proton pairs assignable to the OH^- groups at the hydrophilic ends of opposite lipid layers can also be connected by triplets of (color) magnetic flux

tubes giving rise to the dipole-dipole interaction. This connection need not be permanent and could disappear or appear by the reconnection of the magnetic flux tubes. This could correspond to the transiotion to singlet state for proton pairs and would require energy. The working hypothesis of [17] indeed is that during topological quantum computation the connection is split so that the cell is isolated from external world. The connection would be restored as the computation halts. Photon emission would therefore be seen as a signature of topological quantum computation.

The fact that the proton cyclotron frequency 300 Hz in $B_{end} = .2$ Gauss is the only cyclotron frequency above EEG range, one can ask whether biologically important dark ions form cyclotron Bose-Einstein condensates (possibly also Cooper pairs if fermions), dark protons form a cell membrane spin network, and dark electrons arrange to dark Cooper pairs making cell membrane a super-conductor. This would provide a unified picture about the role of various particle in TGD inspired vision about living matter.

2.2.3 Correlation of photon emissions with the weakening of the Earth's magnetic field

Authors say *During brief periods of imagining white light the power density of photon emissions from the right hemisphere was about* 10^{-11} Wm^{-2} *that was congruent with magnetic energy within the volume associated with a diminishment of* ~ 7 nT *as predicted by the dipole-dipole coupling relation across the neuronal cell membrane.*

The experiment is to some extent a replication of earlier experiment of [4] in which it was observed that visible photon emissions mainly from the right hemisphere is accompanied by a weakening of the horizontal component of the Earth's magnetic field. Decreases over 10 to 15 s of 15 nT and 5 nT at 0.25 m and 1 m from the right side of the head of the subject person were associated with the same magnitude of energy (10^{-11} J) that was associated with the net increase in photon emissions during that period. This energy - assuming each action potential is associated with energy of $eV_{rest} = 1.9 \times 10^{-20}$ J - would be the equivalent of the activity of about 1 billion neurons.

1. If I have understood correctly, the weaking of the magnetic field outside the head of the subject person would be due to magnetic energy change associated with the spin flips taking place in the cell membrane and absorbing the needed energy from this magnetic field. This would obviously represent a new kind of metabolic activity: magnetic field would provide the needed metabolic energy instead of ATP-ADP process. That magnetic body could directly use its magnetic energy to control biological processes, would mean quite a dramatic modification of the usual view about metabolism.

2. The nuclear magnetization disappears for a moment in a transition from spin triplet to spin singlet state, which then spontaneously decay to triplet state again. The excitation of singlet state requires energy so that the magnetic field outside should weaken if it pays the energy bill. The contribution of magnetic dipoles to the horizontal magnetic field component measured outside the head of the subject person disappears and if the direction of dipole magnetization correlates with the direction of the magnetic field the strength of the magnetic field is reduced. The correlation would guarantee that the magnetic fields from different pairs of dipoles do not interfere to zero. Some kind of ordering of the orientations of neurons perhaps induced by the layered structure of cortex and of the almost collinearity of the myelinated axons of white matter is required.

3. Spin-flip transition from triplet to singlet state would change the contribution of magnetic dipoles to the net magnetic field and thus affect the net magnetic field experienced by a test particle. Could this explain the reduction of B_E by factor about 1.8×10^{-4}? At distance of order .1 meter the dipole field created by proton is very small: by a factor 10^{-21} weaker than the 9 nT field created at distance of $d = 10$ nm. The fields of neurons each containing a contribution of about 10^{14} protons sum up and the estimate is that there are 10^9 active neurons. The resulting net factor of 10^{23} could make possible reduction by 9 nT.

4. Triplet-to-singlet spin flip transition taking its energy from the magnetic field is the interpretation suggested by the experiments. The return to the ground state would liberate this energy as large $\hbar_{eff}$ quanta with energies of visible photons transforming later to ordinary visible photons. Therefore the radiated energy could indeed be magnetic energy also in TGD. Of course, also metabolism might drive particles directly to the excited cyclotron states and is expected provide the energy needed to regenerate the magnetic fields since the energy of visible photons is lost.

5. In TGD Universe the correlation of the photon emission with changes (about 7 nT) in the measured magnetic field identified as the Earth's magnetic field B_E having nominal value of $.5 \times 10^{-4}$ T does not force to assign dark photons with the magnetic flux tubes of the Earth's magnetic field.

   (a) One can of question the assignment of 7 nT weakening to B_E as a Maxwellian description not applying in TGD framework. The changes of the horizontal component of the magnetic field are detected outside the head of the subject person is it possible to assign this change to any particular magnetic field? How to distinguish between magnetic fields associated with different space-time sheets? TGD predicts that test particles "feel" their sum if these magnetic space-time sheet have projection in the same region of Minkowski space.

   The possibility to move the flux tubes in such a manner that only the flux quanta of one particular component of the many-sheeted magnetic field contribute to the projection, would allow to analyze the field into these components. Note that un Maxwell's theory this is not possible. The change in the measured magnetic field could be induced by a flux tube carrying 7 nT field assignable to the proton spin network and having a projection to the same M^4 volume as a flux tube of the Earth's magnetic field or the endogenous magnetic field has. Therefore it might not be easy to distinguish between changes of B_E and B_{end}.

   (b) The experimental findings of Blackman et al [3] about the effects of ELF frequencies on vertebrate brain however encourages an interpretation in terms of cyclotron frequencies for magnetic field in "dark" endogenous magnetic field $B_{end} \simeq 2B_E/5$ (this predicts that Ca^{++} cyclotron frequency is 15 Hz, which is not far from 17 Hz). It is of course possible that the flux tubes of the Earth's magnetic field thicken inside the brain so that the strength of the magnetic field is reduced accordingly.

2.2.4 Can one understand the ELF frequencies involved?

Authors state: *Spectral analyses showed maxima in power from electroencephalographic activity within the parahippocampal region and photon emissions from the right hemisphere with shared phase modulations equivalent to about 20 ms.*

The time scale of 20 ms appears also in the experiments of articles 2 and 3 in which rotating and frequency modulated magnetic fields where applied. This time scale corresponds to 50 Hz frequency, which has been found to have biological effects [10]. The cyclotron frequency of Lithium (bosonic ion) for $B_{end} = .2$ Gauss equals to 50.1 Hz (see the appendix of appendix of [13]).

Authors continue : *Beat frequencies (6 Hz) between peak power in photon (17 Hz) and brain (11 Hz) amplitude fluctuations during imagining light were equivalent to energy differences within the visible wavelength that were identical to the intrinsic 8 Hz rhythmic variations of neurons within the parahippocampal gyrus.*

Can one understand the ELF frequencies involved? In TGD framework [13] cyclotron states of electrons, protons, and of ions are possible [13].

1. Ca^{++} is one important bosonic ion able to form cyclotron Bose-Einstein condensates and the 17 Hz frequency for the power of photon fluctuations could correspond to $f(Ca^{++}) = 15$ Hz: note that the strength of the endogenous magnetic field is expected to be under homeostatic control and thus vary in some range.

2. 11 Hz frequency is perhaps too far from alpha frequency 10 Hz but rather near to cyclotron frequency 11.4 Hz for Mn^{++} or 10.8 Hz of Fe^{++} in the field $B_{end} = .2$ Gauss (see the appendix of appendix of [13]).

3. The superposition of effects on test charges caused by MEs associated with 17 Hz and 11 Hz frequencies would give 6 Hz beat frequency. Note that K^+ and Cl^- (fermionic ions) have cyclotron frequencies 7.5 Hz and 8.5 Hz and their Cooper pairs might relate to parahippocampal 8 Hz frequency.

3 Second article

Second article has the title *Demonstration of Entanglement of Pure Photon Emissions at Two Locations That Share Specific Configurations of Magnetic Fields: Implications for Translocation of Consciousness.*

In the article [6] the group reports an excess correlation between "pure" photon emissions at two locations separated by few meters that share specific correlations of frequency modulated magnetic fields. The photon emissions were from chemical reactions. The abstract of the article is following.

The experimental demonstration of non-locality for photon emissions has become relevant because biophotons are coupled to conscious activity and cognition. The experimental condition that produces doubling of photon emissions from two loci during simultaneous chemical reactions when exposed to a sequence of circular rotating magnetic fields with differential phase and group angular velocities was applied to photons from LEDs (light-emitting diodes). A significant but weaker enhancement of photon emissions as measured by photomultiplier tubes occurred when the two LEDs were activated simultaneously within two loci separated by several meters. The effect suggests that under optimal conditions photons emitted from two, magnetic field congruent, loci become macroscopically entangled and that the two loci display properties of a single space. Implications for the transposition of consciousness over large distances are considered.

What was observed was enhanced visible photon emission from of LEDs subject to the same magnetic stimulation as the cell culture dish (neurons) in the earlier experiment [2]. The size of the effect was however smaller. If the effect is real, the presence of the cell culture dishes is not absolutely necessary for the effect although in enhances it. The conclusion of authors is that photons are carriers of consciousness. TGD inspired interpretation is that the experiment conforms magnetic flux tubes as generators of macroscopic quantum coherence.

3.1 Experimental arrangement and results

The article describes first earlier similar experiment [2] using instead of LEDs chemical reactions occuring in cell culture dishes (neurons) and leading to a doubling of photon emissions serving as a signature for coherence - or entanglement as authors express it. LEDs were motivated by the hypothesis that photon field can be equated with consciusness, and tot est this the cell culture dishes were replaced with LEDs. A weaker but significant enchancement of LED emissions is indeed reported.

In the collowing I shall consider mostly the earlier experiment [2] involving cell culture dishes which is identical to the recent one for the mentioned replacement.

1. The distance between the cell culture dishes was few meters as was also the distance of the solenoids from the sample located circularly around it. If I have understood correctly, the circular arrangements of solenoids were in parallel planes around the cell culture dishes (neurons) and the solenoids were directed radially to the dishes: otherwise it would not be possible to achieve a rotating magnetic field.

2. Each set of eight solenoids in circular arrangement around the cell culture dish received identical patterns of piecewise constant magnetic fields generated by potentials having 8 different values: the

duration of single constant piece was 1 ms. Each solenoid created a magnetic field, whose lines emanating from the end of the solenoid were directed to the center of the cell culture dish.

3. Figure 1 of [6] describes the shapes of the AD (accelerating angular velocity, decreasing "phase" modulation) and DI (decelerating angular velocity, increasing "phase" modulation). AD configuration was represented for 8 minutes and followed by DI configuration induced the effect and it occurred immediately after the initiation of DI phase.

Consider now a more detailed description of the AD and DI phases of magnetic stimulation.

1. During AD phase the accelerated rotation of the magnetic field was achieved by creating a magnetic pulse of duration 20,18, 16,..., $T_n = 20-2n, ...$ ms to subsequent solenoids so that only single solenoid contributed to the net magnetic field at any moment. This series was repeated for every rotation of 2π. During AD phase the frequency modulation was slowed down meaning the frequency decreased and also this process was same for every rotation of 2π. The optimal duration of AD phase was about 4-5 minutes.

2. During DI phase decelerated rotation was achieved by in increasing the subsequent durations by 2 ms so that a series of pulses with durations 18 ,20, 22,...,$T_n = 20 + 2n, ...$ ms was obtained. During this period frequency modulation was increased.

3. What "frequency modulation of phase" precisely means? Pictures of AD and DI temporal patterns of voltages (equivalently magnetic fields) feeded to the solenoids inducing a series of values of magnetic field are given Fig. 2 of [6].

   A more detailed description can be found from the earlier article by Persinger's group [2]. The voltage range [-5 V, 5V] was discretized to 8 pieces and the possible discretized voltages in this range are represented by 8 bits. The bit patterns were selected so that they were "physiologically patterned". The value of the magnetic field inside solenoid for n:th bit was proportional to V_n. The duration of each voltage was 1 ms - basic frequency of brain synchrony.

   During AD pattern a) with decreasing frequency and during DI pattern b) with inreasing frequency was used. The numbers of points which composed each pattern were 859 (duration was 859 ms) for AD and 230 (duration of 230 ms) for DI. Only a part of the pattern could be represented since the duration of single 2π rotation was 104 ms, which corresponds to 10 Hz, a fundamental bio-rhythm (Unless there was scaling of the bit duration).

4. Within the center of the 8-solenoid configuration the value of the magnetic field averages to 1 μT. A natural assumption that this magnetic field contributes to the net effective value of the endogenous magnetic field B_{end} inducing small variations of B_{end} in turn modulating cyclotron frequencies.

   The modulated cyclotron frequency should be higher that frequency of modulation and thus higher than 1 kHz. For B_{end} this leaves only electron with cyclotron frequency $f_c = 6 \times 10^5$ Hz under consideration. The effect would be on electron Cooper pairs in the case of cell culture dishes or electrons in the case of LEDs. Electrons are indeed essential also for the function of LED.

5. The frequencies $f_n = 1/T_n$ defined by the durations of magnetic field vary during AD phase between 50 Hz and and 157 Hz. During DI phase the frequencies vary between 50 Hz and 30 Hz. In [10] it is reported that 50 Hz frequencies have biological effects. As already noticed, 50.1 Hz corresponds to cyclotron frequency for Lithium (bosonic ion) for $B_{end} = .2$ Gauss.

3.2 Reconnection of magnetic flux tubes as a mechanism generating macroscopic quantum coherence

A doubling of the rate of emissions of visible photons immediately after the AD phase in the earlier experiment [2] and weaker enhancement in the recent experiment using LEDs instead of cell culture disches, is interpreted as a signature of entanglement. Quantum coherence is perhaps a more appropriate manner to express the findings of the two experiments although quantum coherence makes possible also quantum entanglement. To my opinion the experiments provide support for the basic prediction of TGD inspired quantum biology that magnetic flux tubes are generators of macroscopic quantum coherence.

What seems necessary is that some flux tubes emanating from the solenoids must reconnect to form flux tubes connecting the two cell culture dishes or LEDs: reconnection is indeed one of the fundamental processes in TGD inspired theory of living matter. Without reconnection the flux tubes of the two magnetic fields remain disjoint and cannot induce macroscopic quantum coherence. The reconnection can occur only if the temporal and spatial patterns of the rotating and modulated magnetic fields are identical. These flux tube connections would induce quantum coherence by effectively binding the two systems to single system.

The doubling of the photon emission rate in the earlier experiment involving cell culture can be understood by the well-known rule that in incoherent emission the total rate is N times the individual rate, and in coherent emission N^2 times the individual rate: now N equals to 2. Also destructive interference becomes possible when the summed amplitudes are in opposite phases. This would reduce the rate below the predicted based on incoherence.

Also the enhancement of the photon emission rate from LEDs in a similar arrangement supports the view that macroscopic quantum coherence generated by the magnetic field patterns is relevant and implies that the amplitudes describing the emission of photons from the two LEDs add coherently with some probability so that constructive or possibly also destructive interference occurs. To make this statement more precise, one would need a detailed quantum model for LEDs.

3.3 Why AD followed by DI is needed to induce enhanced photon emissions?

Why should AD period followed by DI period be most effective in inducing photon emissions? Why the flux quanta (flux tubes) do not induce any effects, when the angular velocity is constant and frequency is absent (constant magnetic field)?

1. Accelerated rotation during AD period corresponds at quantum level to an application of magnetic flux tubes from directions $\phi_n = n \times 2\pi/8$ such that the duration of the pulse is reduced in discrete steps. The process should generate frequencies coming as harmonics of $f_n = 1/T_n$. The patterns of magnetic field consisting of periods of constant magnetic field lasting 1 ms and fixed for AD and DI to be "physiologically patterned" determines the Fourier decomposition. The duration of 1 ms brings in harmonics of kHz resonance frequency.

2. The variation of the duration of the magnetic field makes possible to scan a wide range of resonance frequencies of the cell culture. The process would be like tuning a radio. At special frequencies resonant coupling to the frequency of magnetic field and to the frequency defined by the duration of magnetic field becomes possible and enhanced dark photon emissions take place. If the fundamental frequency were not varied, the effect would occur only for very special pulse durations.

3. Why the visible photons were observed only during the beginning of DI phase? If the emitted photons were dark having very long wave length but energy of visible photon, they would not have been detected during AD phase. The decay of dark photons after the beginning of DI phase to bunches of ordinary photons could explain the observed enhanced emissions of visible photons.

3.4 Why the magnetic pulses from a given direction arrived with frequency of 10 Hz?

The magnetic pulses arriving from a given direction to the cell culture dish/LED came with frequency 10 Hz. That a fundamental biorhythm is in question, cannot be an accident. In TGD framework 10 Hz frequency corresponds to the secondary p-adic time scale assignable to electron and defines the size scale of causal diamond assigned with electron. This conforms with the assumption that electronic Cooper pairs are fundamental for consciousness serving also as carriers of super-current through cell membrane. In fact, all elementary particles correspond in zero energy ontology to macroscopic time scales via the secondary p-adic time scales associated with them and for quarks the time scales correspond to frequencies of order 10 ms.

4 Third article

Third article has the title *Experimental Demonstration of Potential Entanglement of Brain Activity over 300 Km for Pairs of Subjects Sharing the Same Circular Rotating, Angular Accelerating Magnetic Fields: Verification by s_LORETA, QEEG Measurements.*

In the third article [7] the group reports excess correlation of brain activity of subject persons separated by 300 km and sharing the same circular rotating, angular accelerating magnetic fields. The abstract of the article is following.

In order to test the presence of excess correlation, or entanglement, pairs of subjects separated by 300 km were either exposed or not exposed to specific configurations of circular magnetic fields with changing angular velocities that dissociated the phase and group components. When one person in the pair was exposed to sound pulses but not to light flash frequencies within the classical electroencephalographic band, there were discrete changes in power within the cerebral space of the other person even though they were not aware of the stimulus times and separated by 300 km. The intra-cerebral changes that only occurred if the magnetic fields were activated around the two cerebrums simultaneously were discrete and involved about single, punctate volumes of about 0.13 cc (125 mm^3*). The potential energy from the applied magnetic field within this volume was calculated to be about* 6×10^{-14} *J and with an average brain power frequency of 10 Hz would result in* 6×10^{-13} *W. Assuming* $\pi \cdot 10^{-2}$ *m*^{2} *for the surface area of the cerebrum, this is equivalent to* $\sim 2 \cdot 10^{-11}$ *Wm*$^{-2}$*. This power density is the same order of magnitude as that associated with photon emission during cognition. Given the average of* 6×10^6 *neurons per 125 mm*^{3}*, the induced energy is equivalent to about* 10^{-20} *J per neuron. This value can be considered a quantum of universal energy and would be congruent with a condition that could promote non-locality.*

4.1 Experimental arrangement and results

If I have understood correctly, the experimental arrangement was roughly following.

1. Two subject persons were involved. Second subject was 300 km away. The other subject person received stimuli at various frequencies of sound or flashes of light while the other person was unaware of these representations. Both members of the pair were exposed to a rotating, circular magnetic field whose frequency modulation would vary with rotation angle. This guarantees that the phase and group velocities of the magnetic field varied and were different.

2. It seems safe to assume that the magnetic field pattern used to stimulate brains of subject persons was identical with that applied in the second experiment.

Authors report a correlation between subject persons in the sense that there discrete changes in EEG power with the cerebral space of the other person even if he/she was not aware of the stimulus times. The effect occurred only if the phase and group velocities assignable to the magnetic field were different. Authors interpret this as entanglement identified as excess correlation if the fields were activated around cerebrum simultaneously and were discrete and involved about single punctuate volumes of about 125 mm^3. Entanglement in this sense need not correspond to quantum entanglement although it could make it possible.

Authors introduce what they call quantum universal energy $E = 10^{-20}$ J, and estimate the that this is the induced energy per neuron transferred from the magnetic field to energy of EEG. In particle physicist's units this gives $E = 6.24 \times 10^{-2}$ eV. This would naturally correspond to energy gained by electron or proton in the resting potential E_{rest}, which is above $E_{min} = 6.15 \times 10^{-2}$ eV. Note that threshold potential for nerve pulse generation corresponds to energy $E_{thr} = 5.5 \times 10^{-2}$ eV. On the other hand, also the first experiment and predecessor of the second experiment involved visible photon emissions which suggests that also visible photons were emitted and they came from the transitions of the proton spin network associated with cell membrane proposed by Wu and Hu [8].

4.2 TGD based interpretation

TGD interpretation should rely on the notion of magnetic body and a model for neuronal membrane as a super-conductor - at least electronic but possibly also ionic super-conductor), cyclotron Bose-Einstein condensed of biologically important ions, and the spin network of dark protons associated with the cell membrane discussed in TGD based model for the outcome of the experiment described in the first article.

1. The flux tubes of the rotating magnetic field would connect the subject persons to single coherent unit reacting to the stimuli posed on second subject like single unit. TGD assigns to the magnetic bodies large effective value of Planck constant so that photons with energies of order E would correspond to much longer wavelengths essential for the coherence in scales of order few wave lengths.

2. The wave length $\lambda = 300$ km could correspond to the Planck constant $\hbar_{eff} \simeq \lambda/\lambda_0 = 1.5 \times 10^{10} \times \hbar$, where one has $\lambda_0 = c/E\hbar \simeq 20$ μm is the wavelength of photon with "quantum universal energy". This energy is in IR region just around thermal threshold. The corresponding period and frequency are $T = c/\lambda = 1$ ms and $f = 1$ kHz, which correspond to fundamental time scales for cell membrane with 1 ms defining the time scale of nerve pulse and 1 kHz defining an important resonance frequency in brain associated with the generation of coherence. Probably this is not an accident. The authors indeed mention that the effect is maximal at distance of 300 km.

Concerning the detailed interpretation of the experiment there are several options. First, TGD suggests two alternative models for cell membrane as Josephson junction involving currents of electron Cooper pairs and possibly also bosonic ions or Cooper pairs of fermionic ions. For the conservative option the cell membrane would be far from vacuum extremal carrying strong induced Kähler field. For the non-conservative option the cell membrane would be nearly vacuum extremal making it maximally sensitive to sensory input. Secondly, the universal quantum suggests emission of dark IR photons, whereas the emission of visible photons associated with cognition suggests visible photons.

1. The quantum universal energy $E = eV_{rest} = 6.24 \times 10^{-2}$ eV would naturally correspond to the energy gained by electron or proton in a membrane potential slightly above the threshold potential. Also the conservative option for cell membrane as Josephson junction would predict Josephson radiation emitted at multiples of Josephson frequency $E = eV_{rest}$ or $E = eE_{thr}$.

2. The non-conservative option for the cell membrane as Josephson junction predicts that the emitted photons have visible energies. This option might be realized for photoreceptors in retina, which react to the sensory stimulus by variation of membrane potential instead of nerve pulse. The correlation of cognition with the emission of visible photons allows also to consider the possibility that some neurons are near-to vacuum extremals (also glial cells as cells which do not generate nerve pulses could be such). Since visible photon emissions are mostly from the right hemisphere, one can ask whether the emissions from the left hemisphere are in IR region and those from right hemisphere in visible region and whether the different ground states of neurons as far-from *resp.* near-to vacuum extremals could distinguish between right and left hemisphere.

3. How does the spin network model based on dark proton strings relate to this? Since the photons have biological functions, the energies of all kinds of EEG photons should be in the same region of

spectrum: visible or IR for a given hemisphere. For near-to vacuum extremals the argument of Hu and Wu would be modified by replacing ordinary magnetic field with a combination of Z^0 magnetic field and ordinary magnetic field. This would imply that the energy scale would increase just as it does when Z^0 electric field dominates over em electric field. Therefore also the photons emitted by spin network at the right hemisphere would be dark EEG photons with energies of visible photons.

4. An alternative interpretation encouraged by the photon emission associated with cognition is that λ_0 corresponds to the energy of visible photon resulting in the transformation of dark ELF photon produced in the triplet-to-singlet transition of proton pair associated with the cell membrane as described in the interpretation of the first experiment. For a photon with energy 1.77 eV at the red end of visible spectrum this would give $\hbar_{eff} = 4.3 \times 10^{11}$. Interestingly, Cyril Smith [1] reports on basis of his own experimentation that the transformation of low energy photons to high energy photons and vice versa takes place for frequency ratio $f_h/f_l = 2 \times 10^{11}$: the interpretation would be also in this case in terms of $\hbar_{eff}$ [21].

5 Conclusions

The results of the experiments of Persinger et al can be understood in the framework of TGD and the findings allow to develop a more precise view about the role of dark electrons, protons, and ions in TGD inspired quantum biology.

1. The identification of the magnetic flux quanta connecting two systems as generators of macroscopic quantum coherence finds experimental support.

2. The proposal of Hu and Wu about proton spin networks associated with cell membrane has a TGD counterpart in terms of dark proton strings allowing interpretation as dark DNA. The spin-paired protons are assigned to the hydro-philic ends of the two lipids in the layers of the cell membrane and the dark proton strings define an analog of DNA double strand. The model of Wu and Hu is subject to the same objections as the model for cyclotron Bose-Einstein condensates and is circumvented by introducing the hierarchy of effective Planck constants.

3. The fact that photon emissions are detected only from the right hemisphere suggests that both options for the cell membrane as Josephson junction are realized: far-from vacuum extremal option for the neurons of the left hemisphere with emissions in infrared and near-to vacuum extremal for the neurons of the right hemisphere.

To sum up, the resulting framework allows an overall view about the roles of both dark electrons, dark protons, and dark ions in quantum biology according to TGD.

References

Biology and Neurocience

[1] C. Smith. *Learning From Water, A Possible Quantum Computing Medium.* CHAOS, 2001.

[2] M. A. Persinger B. T. Dotta. Doubling of local photon emissions when two simultaneous, spatially-separated, chemiluminescent reactions share the same magnetic field configurations. *Journal of Biophysical Chemistry.* `http://core.kmi.open.ac.uk/display/5850998`, 3(1), 2012.

[3] C. F. Blackman. *Effect of Electrical and Magnetic Fields on the Nervous System*, pages 331–355. Plenum, New York, 1994.

[4] Hunter et al. Cerebral dynamics and discrete energy changes in the personal environment during intuitive-like states and perceptions. *Journal of Consciousness Exploration & Research*. *http://jcer.com/index.php/jcj/article/view/116*, 1(9):1179–1197, 2010.

[5] M. Persinger et al. Congruence of Energies for Cerebral Photon Emissions, Quantitative EEG Activities and 5 nT Changes in the Proximal Geomagnetic Field Support Spin-based Hypothesis of Consciousness. *Journal of Consciousness Expolaration & Research*. *http://jcer.com/index.php/jcj/article/view/277*, 2013.

[6] M. Persinger et al. Demonstration of Entanglement of Pure Photon Emissions at Two Locations That Share Specific Configurations of Magnetic Fields: Implications for Translocation of Consciousness. *Journal of Consciousness Expolaration & Research*. *http://jcer.com/index.php/jcj/article/view/278*, 2013.

[7] M. Persinger et al. Experimental Demonstration of Potential Entanglement of Brain Activity Over 300 Km for Pairs of Subjects Sharing the Same Circular Rotating, Angular Accelerating Magnetic Fields: Verification by s_LORETA, QEEG Measurements. *Journal of Consciousness Expolaration & Research*. *http://jcer.com/index.php/jcj/article/view/279*, 2013.

[8] H. Hu and M. Wu. Action Potential Modulation of Neural Spin Networks Suggests Possible Role of Spin. *NeuroQuantology*. *http://cogprints.org/3458/1/SpinRole.pdf*, (4):309–317, 2004.

[9] H. Hu and M. Wu. Thinking outside the box: the essence and implications of quantum entanglement. *NeuroQuantology*, 5:5–16, 2006.

[10] L. Sidorov and K. W. Chen. An Biophysical Mechanisms of Genetic Regulation: Is There a Link to Mind-Body Healing? *DNA Decipher Journal*, 2(2):177–205, 2012.

Books and articles related to TGD

[11] M. Pitkänen. About Nature of Time. In *TGD Inspired Theory of Consciousness*. Onlinebook. http://tgdtheory.com/public_html/tgdconsc/tgdconsc.html#timenature, 2006.

[12] M. Pitkänen. About the New Physics Behind Qualia. In *Quantum Hardware of Living Matter*. Onlinebook. http://tgdtheory.com/public_html/bioware/bioware.html#newphys, 2006.

[13] M. Pitkänen. Bio-Systems as Super-Conductors: part II. In *Quantum Hardware of Living Matter*. Onlinebook. http://tgdtheory.com/public_html/bioware/bioware.html#superc2, 2006.

[14] M. Pitkänen. Conscious Information and Intelligence. In *TGD Inspired Theory of Consciousness*. Onlinebook. http://tgdtheory.com/public_html/tgdconsc/tgdconsc.html#intsysc, 2006.

[15] M. Pitkänen. Dark Forces and Living Matter. In *p-Adic Length Scale Hypothesis and Dark Matter Hierarchy*. Onlinebook. http://tgdtheory.com/public_html/paddark/paddark.html#darkforces, 2006.

[16] M. Pitkänen. Dark Matter Hierarchy and Hierarchy of EEGs. In *TGD and EEG*. Onlinebook. http://tgdtheory.com/public_html/tgdeeg/tgdeeg.html#eegdark, 2006.

[17] M. Pitkänen. DNA as Topological Quantum Computer. In *Genes and Memes*. Onlinebook. http://tgdtheory.com/public_html/genememe/genememe.html#dnatqc, 2006.

[18] M. Pitkänen. Does TGD Predict the Spectrum of Planck Constants? In *Towards M-Matrix*. Onlinebook. http://tgdtheory.com/public_html/tgdquant/tgdquant.html#Planck, 2006.

[19] M. Pitkänen. General Ideas about Many-Sheeted Space-Time: Part I. In *Physics in Many-Sheeted Space-Time.* Onlinebook. http://tgdtheory.com/public_html/tgdclass/tgdclass.html#topcond, 2006.

[20] M. Pitkänen. General Ideas about Many-Sheeted Space-Time: Part II. In *Physics in Many-Sheeted Space-Time.* Onlinebook. http://tgdtheory.com/public_html/tgdclass/tgdclass.html#newviews, 2006.

[21] M. Pitkänen. Homeopathy in Many-Sheeted Space-Time. In *Bio-Systems as Conscious Holograms.* Onlinebook. http://tgdtheory.com/public_html/hologram/hologram.html#homeoc, 2006.

[22] M. Pitkänen. Nuclear String Hypothesis. In *p-Adic Length Scale Hypothesis and Dark Matter Hierarchy.* Onlinebook. http://tgdtheory.com/public_html/paddark/paddark.html#nuclstring, 2006.

[23] M. Pitkänen. p-Adic Physics as Physics of Cognition and Intention. In *TGD Inspired Theory of Consciousness.* Onlinebook. http://tgdtheory.com/public_html/tgdconsc/tgdconsc.html#cognic, 2006.

[24] M. Pitkänen. p-Adic Physics: Physical Ideas. In *TGD as a Generalized Number Theory.* Onlinebook. http://tgdtheory.com/public_html/tgdnumber/tgdnumber.html#phblocks, 2006.

[25] M. Pitkänen. Quantum Model for Nerve Pulse. In *TGD and EEG.* Onlinebook. http://tgdtheory.com/public_html//tgdeeg/tgdeeg/tgdeeg.html#pulse, 2006.

[26] M. Pitkänen. TGD as a Generalized Number Theory: p-Adicization Program. In *TGD as a Generalized Number Theory.* Onlinebook. http://tgdtheory.com/public_html/tgdnumber/tgdnumber.html#visiona, 2006.

[27] M. Pitkänen. Wormhole Magnetic Fields. In *Quantum Hardware of Living Matter.* Onlinebook. http://tgdtheory.com/public_html/bioware/bioware.html#wormc, 2006.

[28] M. Pitkänen. What p-Adic Icosahedron Could Mean? And What about p-Adic Manifold? In *TGD as a Generalized Number Theory.* Onlinebook. http://tgdtheory.com/public_html/tgdnumber/tgdnumber.html#picosahedron, 2013.

[29] M. Pitkänen. Two attempts to understand PK. http://tgdtheory.com/public_html/articles/PKoptions.pdf, 2012.

Research Essay

Interactionism Read Anew: A Proposal Concerning Phenomenal Judgments

Einar L. Halvorsen*

ABSTRACT

From a classical Cartesian perspective, interactionism implies the transfer of thoughts and feelings from a non-physical phenomenal consciousness to the physical brain. Thereby, phenomenal consciousness is thought to control the physical body somehow like a marionette hanging by strings of non-physical thought. Differing from this depiction, a basic premise of the current interactionist hypothesis is that the non-physical phenomenal consciousness reflexively effects accentuation of thoughts, feelings and sensory experiences which already exist as physical brain processes. In this essay, the mentioned interactionist hypothesis is presented and central philosophical problems which pertain to it are discussed. Conclusively, the hypothesis may withstand the initial scrutiny and is thereby rendered coherent. Nonetheless, the feasibility of interactionism is not thereby significantly influenced. That would require a much more extensive treatment of the feasibility of the current hypothesis as well as of coherent solutions to numerous other problems pertaining to interactionism.

Key Words: interactionism, phenomenal judgment, phenomenal consciousness, physical brain.

Introduction

Like Michael Tye and others (Tye, p. 1), I will refer to phenomenal consciousness as "P-consciousness". The term "P-consciousness" will here provide for an intently undefined meaning, as justifiable in light of Ned Block's notion that the meaning of that term is lingually indefinable (Tye, p. 1). Tye, freely re-formulated, sees P-consciousness as a capacity to possess a phenomenal side to experiences (Tye, p. 144-5). If Block is right, however, the phrase "phenomenal side to experiences", if understood as the indefinable "P-side", makes that description circular.

The assumption of the lingual indefinability of P-consciousness carries implications also for an interactionist perspective. While the physical world constitutes one pole of a purported interactionist dualism, the nature of the other, non-physical pole will remain unknown or a mere object of beliefs. Beyond the notion that P-consciousness is intimately linked to experience, furthermore, a consensual nature of such beliefs can only be assumed.

Interactionism, the general theory discussed in this essay, is the idea that a non-physical P-consciousness causally interacts with the physical brain (Chalmers, 2010, p. 126). Given that the doctrine of interactionism is correct, violation of the doctrine of causal closure of the physical

*Correspondence: Einar L. Halvorsen, Independent Researcher, Germany. Email: EinarLautenHalvorsen@gmail.com

world (Stoljar, p. 220-2), including the quantum mechanical version of that doctrine (Papineau & Selina, p. 74-5), is necessary.

Stewart Goetz convincingly argues that no theoretical hindrance exists which makes scientifically incoherent the notion that causal closure may be violated (Goetz, in Baker & Goetz, p. 104-16). Robin Collins, furthermore, explicates that it is not scientifically necessitated that violation of causal closure must imply violation of energy conservation (Collins, in Baker & Goetz, p. 124-33).

Even if the doctrine of causal closure could coherently be violated, however, another problem is that expressed by the question of how quantum level interactionist influences may amount to neuro-functional changes. This would be necessary should interactionism make a relevant difference (Chalmers, 2010, 126-7n). Sir John C. Eccles early described the neuronal processes which would have to be influenced (Popper & Eccles, p. 232).

A multitude of electro-chemical variables causally co-determine the firing of "action potentials", the single wave-like signals of invariable amplitude which "shoot" through single neurons (Klinke & Silbernagl, p. 51-77). The challenge, if wishing to rule out interactionism, is to prove each such electro-chemical variable as fully determined by reductive processes. David Chalmers suggests that scientific understanding of these matters is still incomplete (Chalmers, 2010, p. 126-7n).

In order that an interactionist solution to the problem of phenomenal judgements may be correct, a necessary premise is the coherence of the notion that *all* further problems with interactionism may be solved. Only two further problems, however, which have direct relevance to the problem of phenomenal judgements as viewed from an interactionist perspective, will here be granted short preliminary treatments.

Functionality and causality

A well known doctrine is that P-consciousness is functionally indefinable (Chalmers, 1996, p. 46-7). A problem is thus whether a non-physical P-consciousness may be the source of functional phenomena within the physical brain. Chalmers argues that if any interactionist influence would exist, that would require non-physical, yet *functional* variables; "psychons" (Chalmers, 1996, p. 156-8), as indeed proposed within Eccles' interactionist perspective (Eccles, p. 87-8).

Thereby, however, the problem is merely displaced from the physical world to a non-physical sphere. The next question becomes whether the functionally indefinable P-consciousness can be the source of the psychons' functioning. From an interactionist perspective, the challenge is to explain how P-consciousness *generally* can be the source of functional phenomena without itself being functionally definable.
As initial analogy, P-consciousness may be compared to an invariant light source. This light could be reflected in the shattered mirror surface of a parabolic sphere (here an analogy to a brain). Those mirror shards retaining an "angle" similar to that of an unbroken parabolic mirror will have surfaces being the most saturated with the light being reflected in them. Other shards,

which surfaces have become more or less parallel to the light beams, will see their surfaces less saturated with light.

While the light itself here symbolizes P-consciousness, the optimal saturation of light on shard surfaces symbolizes the interactionist influence upon brain segments. The saturation of light on shards' surfaces, furthermore, will be fully determined by the "angle" of each shard, not by variances or differentiations within the light source. Hence, any potential changes to the interactionist influence, although change is not represented by the analogy, appear definable by the functional processes of the physical brain itself.

All analogies are imperfect. An imperfection with the present one is that even light emanating from an invariant source is functionally definable as waves of electromagnetic energy. The following account is lacking regarding the structure of what argument it constitutes from a scientific perspective. My intention, however, is merely to draw attention to an *apparent* conceivability of a solution to the problem.

According to the theory of relativity, an implication of length contractions and time dilations occurring by motion (Sartory, p. 185-200), is that the spatio-temporal universe appears as lacking actual extension if seen from the "perspective" of light (Haisch, p. 95-7). That perspective is principally impossible to acquire through scientific observation, since no observing, physical object may ever behave like a photon of light (Sartory, p. 210).

From quantum physics, it is known that the spatio-temporal location of single photons may be indefinable thanks to so called "superpositions" (Halvorson, in Baker & Goetz p. 142-6). Without manifest location, furthermore, there is no manifestation at all. When photons *are* localizable, distances between them appear, through so called "entanglement", as causing no hindrance for their instantaneous "contact" (Halvorson, in Baker & Goetz p. 146-9).

The latter notion potentially even mirrors relativity theory, as also there, it is as if for photons, space and time lacks reality-defining power. Lee Smolin (Smolin, p. 210) predicted that the theory of relativity may be coherent with quantum physical theory only if one assumes that the universe exists according to a holographic principle (Smolin, p. 169-78).

What at any rate appears indicated is that some "factors" of reality exist beyond spatio-temporal parameters. Hence, the idea that P-consciousness could exist in or as some spatio-temporally indefinable state is coherent. P-consciousness could then potentially affect various brain areas non-functionally.

Given that all functionality is definable within spatio-temporal parameters, purported "psychons" could apparently exist only within a non-physical reality having spatio-temporal characteristics. Curiously, however, in light of a *general* solution to how P-consciousness may affect functional phenomena, even the idea of psychons becomes coherent.

What can be assumed transferred to the brain?

By interactionism, one may oft imagine that the non-physical P-consciousness "has" thoughts and feelings which get transferred to the brain. That would require that normal human experience can take place independently of the brain. As well known, however, if a part of the brain is destroyed, the psychological processes which rely on the functions of the damaged brain areas are disrupted. Consequently, I take as premise concerning normal human experience, that all experiential objects; everything we experience through sensing, thinking and feeling, are represented as neuro-functional brain activity.

I assume the same also when specifying experience as phenomenal (P-) experience. From a first person perspective, arguably, no phenomenal character or feature of experience seem barred from becoming motive of directed actions or verbalizations which obviously depend on neuro-functional brain processes. From a third person perspective, it arguably appears as the only rational alternative to trust first hand report and thus assume reports of blindness, example wise, to also imply the phenomenal lack of vision.

In light of the above, a question is how interactionism may be coherent with neuro-science. As an initial consideration, contrary to the position favoured by Tye (Tye, p. 155-82), it appears at the very least coherent that brain processes may be subconsciously accessible to P-consciousness. Not even subconscious experiences, however, will here be assumed to exist independently of brain activity.

Rather, I will assume that a direct effect of an interactionist influence consists of an *accentuation* of experiential objects already existing as patterns of neuro-functional brain activity. This better fits Eccles' belief concerning inter-neuronal communication, that interactionist influences only may "*modify the probability of vesicular emission of the activated synapses*" (Eccles, p. 77).

Eccles nevertheless favoured the view, contrary to that here favoured, that interactionist influences may transfer discrete experiences (Eccles, p. 71-2). Within both Eccles' view and the present, however, an interactionist influence should alter the frequency of action potentials in those neurons which combined activity make up the overall pattern of neuro-functional brain activity constituting the given experiential object.

An interactionist influence which direct effect is the accentuation of pre-existing brain activity could be constituted by a non-intentional, reflex-like mechanism, like in the previous "mirror analogy". It would not have to constitute any consciously experienced effort. Further, it is also not necessitated that the brain in a direct manner should experience any "transmission" which mediates an interactionist influence. Rather, the influence could take place as a fully subconscious process.

In the above, however, no mechanism has been identified whereby the subconscious accentuation of experiential objects should be detectable *as that* by the brain. Another question is then whether the brain may detect the presence of an interactionist influence through a different mechanism. If neither alternative is possible, we may apparently not explain changes to phenomenal judgements as resulting from interactionist influences. The second alternative, however, may in fact be coherent.

Required for this is a stable correlation between two known variables. Firstly, the amount of absorbed neurotransmitter molecules sufficient for the release of one action potential within a neuron (Klinke & Silbernagl, p. 65-6). Secondly, the amount of neurotransmitters absorbed from the first neuron by an adjacent, neuro-functionally subsequent neuron as a consequence of the action potential of the first neuron (Klinke & Silbernagl, p. 62-5).

We may assume that the suggested type of correlation, which may be uneven as long as stable, can operate despite input to the receptor neuron from "third neurons". What is decisive is the amount of neurotransmitters absorbed from the "first neuron" as a consequence of one of its action potentials. Even within networks of interconnected neurons, we may single out, at least as a theoretical construct, the "linear" effect upon single neurons from other single neurons.

A potential reservation is the idea that synaptic transmission from third neurons could affect the very function of the receptor neuron so that the amount of neurotransmitters absorbed from the first neuron is thereby changed. However, such variance should be lawfully determined and thus principally correctable by theoretical models constructed to take height for effects of third neuron influences.

The suggested type of stable correlation will here be called "SLAN-correlation" (Stable Linear Absorption of Neurotransmitters Correlation). Apart from potential effects of interactionist influences, a mode of functioning according to which the SLAN-correlation exists may be ingrained through evolution as a premise for healthy psychological functioning. That notion is supported by the fact that its violation oft appears to be a mechanism of psychopathology (Laruelle, in Hirsch & Weinberger, p. 365-81).

Hypothetically, we can here imagine that some neuro-functional sequences of neurons within the brain are insulated against input from third neurons. Further, we may imagine excitatory synaptic connections between the neurons of such sequences to occur according to a "chain principle". Further, inhibitory synaptic connections (Klinke & Silbernagl, 67) could operate only between neurons separated by one or more intermediate neurons within the "excitatory chain".

Such "insulated sequences" could potentially be activated exclusively in cases of SLAN-correlation violations. Each neuron of the sequence could be wired for transmissions through both excitatory and inhibitory synapses according to the above outlined principles. The result, by intact SLAN-correlation, could be lacking activation from the first inhibitory neuronal interconnection onward. Inhibitory and excitatory influences might then cancel each other out, as permitted by the principles of neuro-physiology (Klinke & Silbernagl, p. 68).

By violation of the SLAN-correlation, however, excitatory influence may outweigh inhibitory influence in receptor cells receiving inhibitory input, thereby causing those cells' activation. The reason is that in those cases, any accentuating effect of SLAN-correlation violations may accumulate over two or more excitatory synapses but over only one inhibitory synapse.

The above account is theoretically rather than empirically based. Any effort, furthermore, to formally demonstrate the theoretical coherence of the account exceeds the spatial confines of this essay. Nonetheless, I hold the coherence of the account as a theoretical premise for the arguments of this essay. For simplicity, activation of neuro-functional "insulated sequences" which lay un-activated except by SLAN-correlation violations will be called "LINS-activation" (Linear Insulated Neuro-functional Sequence Activation).

It is necessary, principally, to allow that LINS-activation may occur without an interactionist influence (positive error) and that it may be absent despite of an interactionist influence (negative error). More generally stated, it is not logically necessitated that SLAN-correlation violations, which may potentially also occur without interactionist influences, should always lead to LINS-activation.

By schizophrenia, as example, deregulation of the amount of neurotransmitters within single neurons leads to what is *here* termed SLAN-correlation violations (Laruelle, in Hirsch & Weinberger, p. 365-81), albeit not extremely abrupt ones. To suggest that this would lead to LINS-activation would appear unfounded, though the idea that it could occasionally do so is not incoherent.

Fluctuations also of other, even subtler and more instantaneously effective electro-chemical variables could cause so called SLAN-correlation violations, however (Eccles, p. 55-69). Those variables, Eccles described, could conceivably be influenced by quantum level changes. Like also Eccles, Chalmers sees the potential existence of quantum level interactionist influences as coherent (Chalmers, 2010, 126-8).

For theoretical reasons, a statistically high degree of correlation between interactionist influences and LINS-activations will here be assumed. Further, the occurrence of LINS-activations could conceivably get registered by the reflectively conscious function of the physical brain as an experiential phenomenon. Pertaining to phenomenal judgements, thus, the present interactionist account could potentially explain how effects of an interactionist influence may become objects of claims and cognitive beliefs.

Phenomenal Judgements & Interactionism

Phenomenal judgements refer to beliefs concerning the phenomenal character of experience (Chalmers, 1996, p. 173-5). Such beliefs can be expressed by claims such as "colours are mysterious" or "I am phenomenally conscious", example wise. The problem of phenomenal judgements is that of giving an explanation which accounts for the existence and the real nature of such beliefs (Chalmers, 1996, p. 184-6).

The "hard problem" of consciousness, strictly interpreted, is that of explaining the existence of P-consciousness (Chalmers, 2010, p. 3-6), not merely the existence of the *belief* that it exists. The problem of "the explanatory gap", next, is that of why there is an indescribable (strictly assumed phenomenal) side to discrete experiences (Chalmers, p. 1996, p. 47). With strict interpretations of both problems, the problem of the explanatory gap will be integral to the hard problem, since all phenomenal experiences require a phenomenal subject (Strawson, in Freeman, p. 189-91).

The hard problem is relevant to phenomenal judgements only insofar as actual phenomenality constitutes a reason for beliefs about phenomenality. Eliminativists have denied the very existence of P-consciousness (Macpherson, in Freeman, p. 75) and logically thus also that of a strict "hard problem". The problem of phenomenal judgements, then, can potentially (but must not) be answered in concert with an answer to the hard problem.

If remaining "undogmatic" regarding the strictness of the hard problem, P-consciousness appears potentially reducible to any reductively definable property which is capable of solving the problem of phenomenal judgements. Tye holds that there may be a crucial difference between "knowledge by description" and "knowledge by acquaintance". Furthermore, that no part of the former can be "part and parcel" of the latter (Tye, p. 139). Feelings that something is missing to statements about phenomenal qualia, example wise, such as also implied by the explanatory gap, could thus be explainable (Tye, p. 143).

For Tye, in the sense that he is no eliminativist, P-consciousness exists yet *is* a reductive property, namely the brain state(s) for which knowledge by acquaintance takes place (Tye, p. 144-5). If his theoretical framework is correct, however, it serves his case of defending physicalism only insofar as it defends any substance monist view, including variances of property dualism (Chalmers, 1996, p. 124-5). Conceivably, it could explain why our cognitive beliefs *correspond* to our phenomenal experience. If so, the hard problem would remain, even as the problem of phenomenal judgements could appear solved.

The phenomenal judgement, however, that without the problem of phenomenal judgements, the hard problem could still remain, relies on first person knowledge. If one chooses to take first person perspectives upon consciousness seriously, it may arguably appear that there is something to P-consciousness which is not only intellectually unexplainable (a mystery), but something taking on an existential importance (a captivating mystery). The philosophical question of what extent to which first person knowledge should be taken seriously (Taliaferro, in Baker & Goetz, p. 26-40), however, will be left untouched at this point of the discussion.

An interpretation of Tye

For the sake of the subsequent discussions, a short interpretation of Tye's perspective on phenomenal judgements will here be presented. Since Tye sees phenomenal experience and knowledge by acquaintance as synonymous, he understands the implicated explanatory gap between knowledge by acquaintance and by description as that which is assumed integral to the hard problem.

Because knowledge by acquaintance has no part or parcel of knowledge by description, Tye sees the character of all phenomenal experiences as lingually indefinable, like also implied by the well known philosophical notion of inverted qualia (Chalmers, 1996, p. 263-6). Contrary to appearance, this is consistent with Tye's conviction that P-consciousness is definable as a property of the function of the physical brain.

Using the colour "blue" as example, that colour is descriptively definable as a pattern of brain activity. It appears that also the phenomenal character of blue experience is "fixed" by the neuro-functional characteristics of that very same brain state. This should be assumed even by property dualism and interactionism, given that all normal human experience relies on brain activity.

Tye holds that even from a purely reductive, physicalist perspective, one may *recognize* (pick out) "blue" as a colour separate from other colours regardless of whether one is aware of or able to describe its neuro-functional characteristics (Tye, p. 139). Descriptions of the brain states defining the experience, furthermore, may never convey the character of the experience by acquaintance which allows the recognition. We see, thus, that although Tye sees P-consciousness as neuro-functionally definable, he does not believe the same to concern the subjective *character* of discrete phenomenal experiences.

The character of experiences by acquaintance represents "brute qualities", identifiable only by expressions such as "one of those" (Chalmers, 1996, p. 288-92). True, they can also be referred to using some random token name, like "blue". The meaning even of such consensual token names, however, is flexible given the existence of inverted spectrum variances. Further descriptive knowledge will also be unhelpful in specifying the phenomenal character of the experience of a concepts' referent. Example wise, telling someone that blue is the colour of the sky is unhelpful given inverted spectrum variances.

A "concept", Tye holds, is deferential (Tye, p. 40-1), meaning that it can be "possessed" even if it is not fully understood (Tye, p. 63-74). A person thus possessing the concept BLUE without full understanding could say about the colour indigo that it is blue. He or she could be colour blind and experience all blue as indigo. Hence, it is not necessary to have undergone acquaintance with the colour blue in order to possess the concept BLUE (Tye, p. 66).

There is, of course, a problem of defining "full understanding of concepts". True, if a person is either colour blind or experiencing inverted qualia, he or she must logically, at least within an assumed physicalist reality, differ from other people as to how colours are represented in his or her brain. Unless the presence of the relevant type of neuro-functional brain activity is determined in each single case, however, it will remain a matter of mere presumption which phenomenal character is consensually assigned to each colour concept.

The above underlines that, for all practical purposes, Tye presents a physicalist account which respects the lingual indefinability of phenomenal qualia. What might still appear as a remaining question is that of why people commonly fail to realize that the character of experiences by acquaintance cannot be referred to by mere token names of brute qualities.

The principle of inverted qualia might rarely be spontaneously realized, yet this does not make it a scientific mystery. The erroneous belief that concepts may refer to phenomenal qualia appears to rely on a simple ego-centric error, namely the implicit, un-reflected belief that the own subjective perspective is defining of interpersonal reality. In summary, thus, Tye's perspective allows the notion that people believe to refer to phenomenal qualia using "normal" concepts.

A non-physicalist perspective

Initially in agreement with a non-interactionist, property dualistic perspective (Chalmers, 2010, p. 243-4), I suggest that phenomenal experience is conveyed by the same patterns of brain activity which define experiences in cases of knowledge by acquaintance. The phenomenal side to the experience of any experiential object will thus be definable as the phenomenal experience *of* the neuro-functional pattern of brain activity defining the experience of that experiential object. We here end up with a double explanatory gap, one between knowledge by acquaintance and knowledge by description, another between phenomenal and non-phenomenal experience of that known by acquaintance.

A comment on the relationship between knowledge by acquaintance, knowledge by description and the meaning of concepts will be useful for some of the following discussions. Using Chalmers' example, the concept WATER can have different secondary intensions (known a posteriori) like "H_2O" or, in a hypothetical other world; "*XYZ*" (Chalmers, 1996, p. 57). The primary intensions could in both cases be "*the dominant clear, drinkable liquid in the oceans and lakes*" (Chalmers, 1996, p. 57).

However, the brute qualities of water known by acquaintance must possess an a priori nature even relative to its primary intension. Through its descriptive content, the latter is generalizing and classifying of what may (but must not) be a priori "acquaintances" with the brute qualities of the experience of water. Sainsbury and Tye similarly hold that the so called "two-dimensional semantic" with primary and secondary intensions "*fails to connect in a natural way to the notion of a priori knowledge*" (Sainsbury and Tye, p. 36).

From the perspective of an originalist theory of concepts, the acquiring (taking into possession) of a concept can be separate from its origin, since concepts typically are shared (Sainsbury & Tye, p. 40-4). Hence, the acquiring of concepts may oft reflect the adoption of a non-originating use of the given concept, although one may later learn the concept's originating use.

In such cases, acquiring cannot occur based on a priori experience by acquaintance of a non-descriptive referent phenomenon. Even if the originating use of a concept would have such a priori experience as referent, through adopting a non-originating use, the acquiring will not involve such experience. Oft, furthermore, even the originating use of a concept has merely descriptive referents, like in the case of the concept QUARK (Sainsbury & Tye, p. 42-3).

Sometimes, however, the acquiring and the origin of a concept may be synonymous (Sainsbury & Tye, p. 42). This allows the occurrence of discoveries that the given concept has the same originating use as a pre-existing, shared concept (Tye, p. 39-40). It appears that exclusively in such cases, the use of concepts *can* (but must not) reflect the a priori experience of the given concepts' non-descriptive referent phenomena by acquaintance.

Although here assumed that only experiences by acquaintance may have phenomenal sides, even concepts which use involves merely descriptive referents constitute experiential objects existing as patterns of brain activity. Consequently, also these should *secondarily* be possible to experience a priori by acquaintance. Even in the case of a posteriori, descriptive knowledge, thus, it will be assumed that one may possess a phenomenal side to the experience *of* this knowledge.

The core interactionist hypothesis

The non-physical P-consciousness is assumed only *ideally* to be the source of the previously purported LINS-activation. Furthermore, no identified mechanism appears to allow the brain to decide whether the SLAN-correlation is violated because of an interactionist influence or thanks to some reductively definable cause. Indeed, SLAN-correlation violations may not be detectable in any direct way from the perspective of the psyche.

I previously suggested that they may at least sometimes be detected indirectly through the occurrence of the LINS-activation. Another consequence of SLAN-correlation violations could logically be the abrupt, maybe intrusive appearance of thoughts, feelings or sensory impressions to conscious awareness. As an extreme, but thereby obvious example, we could take schizophrenic hallucinations (Laruelle, in Hirsch & Weinberger, p. 365).

As a potential consequence of all such experience, furthermore, the psyche may cumulatively come to experience a lack of autonomy and integrity of the psyche. The apparent truth; that the psyche does not "master" the physical brain, but can only function adaptively given a fine, homeostasis-like balance within it (the SLAN-correlation) may hypothetically constitute an existential threat to the psyche. It is a known doctrine within psycho-analytically oriented theory that the psyche harbours "unrealistic needs", such as the preservation of an illusion of solidity, permanence and perfection, which Charles Hanly described as "*a self-image that is distorted by idealization*" (Epstein, p. 24).

Analogically, a need of the psyche not to experience itself as at the mercy of potentially merciless neuro-physiological processes of the physical brain is conceivable. An archaic and subconscious psychic defence mechanism is "projective identification", the ejection of internal threats in order to perceive (identify) them as external ones (Ogden, p.144-6). Wilfred Bion held that projective identification can account even for many bizarre psychotic experiences of schizophrenics (Ogden, p. 146).

By schizophrenic psychoses, patients frequently claim to have thoughts that are not their own (AMDP, p. 85-6). As above assumed, such "intrusive" experiences may cause a sense of the psyche's lacking autonomy or integrity and be experienced as threatening. Beside the mentioned sense of thought insertion, projective identification may lead to delusions of alien influence (AMDP, p. 86-7) and persecution (AMDP, p. 68-9). The latter could more directly account for the projective identification of an experienced, internal threat.

Also a priori experience by acquaintance of the LINS-activation may, according to the same principle as above outlined, be experienced as a threat to the autonomy and integrity of the psyche. This, because LINS-activations, which is thought to exclusively follow SLAN-correlation violations, may be experienced as abrupt, intrusive presences felt by the psyche as something uncontrollable.

However, if a concept LINS-ACTIVATION would be rightly possessed, unlike typically the case among other concepts having referents being experienced by acquaintance, it might represent a phenomenon with no positively identifiable referent. That is, it may logically represent nothing in the *external* physical world and might also constitute no discrete thought, feeling or illusory sensory impression.

Hypothetically, the LINS-activation could be experienced as was it nothing at all, still penetrantly present *as that* in an a priori fashion. Effectively, thus, the LINS-activation could be unique as being impossible to positively describe a posteriori, bar in terms of the very brain activity constituting its neuro-functional referent. Devoid of any positively identifiable, descriptive referent phenomena, the only accessible referent of the LINS-activation might thus be the very character of its experience by acquaintance.

Descriptively identifiable only as negations like "nothing" or "absence", the LINS-activation may find no physical referent at all. By projective identification, thus, the only accessible referent of the LINS-activation, the character of its experience by acquaintance, might thanks to lacking alternatives be falsely judged as transcendent and ontologically non-physical. Purely hypothetically, illusory beliefs in transcendent qualities to experience may even be "welcomed" by the psyche, since transcendence could serve the "unrealistic needs" of solidity, permanence and perfection.

According to the current hypothesis, beliefs in a non-physical character to experiences would be "falsely founded", since the LINS-activation actually refers to a physical brain phenomenon. Hence, the presumed, *actually* non-physical nature of P-consciousness would not be the direct reason for beliefs about that very non-physical nature. Still, it could constitute an indirect reason for it in the form of an interactionist influence.

Answer to some criticisms

A potential criticism of the above interactionist hypothesis can be expressed through the question of how we justify the idea that phenomenal experience merely is present as the LINS-activation. Answering this, it should once more be clarified that between phenomenal and non-phenomenal experience, no different functionality in the brain is here assumed, like also not in cases of property dualism.

Presumably, there is always a phenomenal side to experiences by acquaintance. The purported LINS-activation entails, on the one hand, one single and separate phenomenal experience among the multitude of other, phenomenal experiences. On the other hand, it may also constitute the reason for convictions about a quality over and above everything physical to the experience of any experiential object currently accentuated by an interactionist influence

One could still ask how, without adding any new experiential contents, the LINS-activation alone may explain judgements about the entire range of phenomenal experiences. The answer is that the LINS-activation is assumed to ideally occur because of an interactionist accentuation of discrete experiences already present. The decisive factor regarding the character of each phenomenal experience is then the pre-existing experience *being* accentuated by the interactionist influence.

The above may still be further questioned. In light of the above accounts, it appears that people should believe recognizing a reductively indefinable character to the experience of the LINS-activation rather than to that of the experiential object being accentuated. A confusion may be comprehensible, however, because the LINS-activation and the accentuation of the present experiential object presumably occur simultaneously and co-dependently upon an interactionist influence.

Further and as previously mentioned, the LINS-activation may be thought of as being, in a sense, "object-less". That is, the only identifiable experiential "contents" could be those of the experiential object *being* accentuated. Given the sum of those circumstances, a fallacy of correlation and causality (Losee, p. 41-8) could conceivably occur. Hence, one might believe the impression of a transcendent, non-physical quality to the character of any present experience (a-phenomenally defined) to be caused by the presently accentuated experiential object rather than by the LINS-activation.

Another question is that of whether the present interactionist account may explain why and how the physical brain can believe in the existence of phenomenality to experience. After all, the brain must produce the claims reflecting the beliefs about phenomenality. The flip side to the same problem is the question of how P-consciousness may believe that functional brain states refer to its own phenomenal experiences.

Even within an interactionist account of reality, however, we must assume that the physical brain alone may know nothing about phenomenal (P-) experience. Like also according to a property dualistic rationale (Chalmers, 1996, p. 124-5), the physical brain and P-consciousness could mutually lack ways to "inform" the other that a discrepancy exists between an experience as functionally definable and its phenomenal side. The discrepancy may simply not matter as long as the neuro-functional determinants of the experience are identical.

Thus, even as one experiences the phenomenal side to an experiential object, one may be "blind" to the fact that there is a difference between an experience as functionally definable and its phenomenal side. This also implies "blindness" to any potential experience that there should be a mystery to the notion that the character of phenomenal experience is reductively definable.

In the absence of any LINS-activation, thus, P-consciousness may experience the phenomenal side to discrete experiences by acquaintance without accompanying beliefs that the character of those experiences is transcending everything physically definable. The specific case of the LINS-activation, however, may constitute an exception. As previously outlined, the character of its experience by acquaintance could hypothetically be experienced as transcendent and ontologically non-physical.

The explanatory power of the present hypothesis

An important question is whether an interactionist influence could result in beliefs differing from those of a phenomenal zombie (Chalmers, 1996, p. 94-9) following "normal" knowledge by acquaintance. In light of the difference between knowledge by acquaintance and by description, zombies' beliefs about a lingually indefinable character to experiences may take place without interactionist influences. Exactly like Tye holds, this could appear to solve the problem of the explanatory gap. This would also leave the present interactionist hypothesis stripped of any unique explanatory power.

Rightly, the character of the experience by acquaintance of the colour blue is seen as indescribable from Tye's perspective. However, within any substance monist view, of which physicalism and variances of property dualism are examples, a strong subjective conviction that the indescribable character of blue experience is transcendent should not arise. As above mentioned, P-consciousness would arguably be unable to detect any major mystery to the notion that phenomenality should be reductively definable.

It was suggested, however, that the experience by acquaintance of the LINS-activation could appear as having a non-physical character to it. Obviously, even without any purported LINS-activation one may form *hypotheses* (and incidentally even favour these) that the character of experiences by acquaintance may best be explained as transcendent. This would then work according to the principles of scientific hypotheses in general, which of nature are a posteriori, matter-of-factly and affect-less.

Violations of the matter-of-factly and hypothetical nature of claims as expectable concerning a posteriori descriptive notions could indicate the presence of the direct a priori experience of some phenomenon. If, additionally, such "irrational convictions" would concern notions of transcendent or ontologically non-physical aspects of experience, this would fit that which is expectable within the current interactionist hypothesis.

As previously mentioned, both positive and negative errors might occur regarding the co-occurrence between an interactionist influence and LINS-activation. Thus, irrational convictions must sometimes be granted to be "just" irrational. Given that errors are exceptional, however, the current interactionist hypothesis generally favours the notion that first person knowledge may have a central role when it comes to our judgements about the nature of consciousness.

A question is still why we should prefer the present interactionist hypothesis over new, more complex forms of substance monism. In those alternative scenarios, the LINS-activation always occurs by reductive means, still causing phenomenal judgements about the character of phenomenal experience as being transcendent. Within a property dualistic scenario, as example, one could imagine that beliefs about phenomenality would fully correspond to *actual* phenomenality, yet the beliefs would not even indirectly be caused by that phenomenal reality.

However, the LINS-activation would then have to be seen as resulting from random deregulation of the SLAN-correlation. This, furthermore, does not appear to be the principle according to which our phenomenal judgements are produced. Example wise, we do not hear claims that the character of the experience of the colour blue *some times* and *without predictability* has a quality transcending everything physically definable.

In summary, although a traditional property dualistic perspective may explain beliefs that one refers to phenomenal experiences, it leaves the subject "blind" to the difference between phenomenal experience and knowledge by acquaintance. Tye's physicalism explains the same beliefs without that dilemma, since it sees phenomenal experience and knowledge by acquaintance as synonymous. Nonetheless, it seems that only an interactionist alternative may account for the mystery of consciousness as expressed by the first person experience of the hard problem.

The problem of lingual reference to P-consciousness

As evident by the fact that people try to express convictions about P-consciousness in language, certain concepts must frequently be believed to refer to P-consciousness. All thus relevant concepts, however, have reductively definable referents, existing independently of any purported interactionist influence. A fitting example is exactly the concept PHENOMENAL CONSCIOUSNESS, here understood as distinct from P-consciousness.

The word "phenomenal" is a term from philosophy, especially linked to 19th century philosopher Edmund Husserl. It concerns the manner in which experiential objects (phenomena) "appear" to consciousness (Fahey, online). The word "consciousness", next, can be defined as a mode of functionality which can be shared by the neuro-physiology of biological organisms and artificial computing systems (Chalmers, 1996, p. 275).

Tye argues that a reductively definable mental state (or states) *for which* knowledge by acquaintance takes place likely is the referent of P-consciousness (Tye, p. 144-5). I also judge it as safe to infer that Tye, from his physicalist perspective, sees that (or those) state(s) as the referent phenomenon of the concept PHENOMENAL CONSCIOUSNESS. For reasons of simplicity, such (a) mental state(s) will here be referred to as "the mental state of acquaintance". In that state, no agency appears involved, only passively "receptive" experience.

Many concepts other than PHENOMENAL CONSCIOUSNESS could have the same, reductively definable referent. Examples are concepts such as PURE BEING and SUBJECTIVITY. Another example is the concept I, which is arguably also used with the belief that it refers to P-consciousness. Accordingly, statements like "my brain registers colours, but only I see their qualia" may follow. According to Mark Epstein (Epstein, p. 47-8), the "I" is a portion of the psychological ego. Further, all psychological phenomena are functionally definable (Chalmers, 1996, p.46-7).

A question is whether the reductive referents of the two concepts PHENOMENAL CONSCIOUSNESS and I could be reduced to one single, neuro-functional brain phenomenon. Given that the two concepts may be believed to refer to the same phenomenon, we would thereby avert a theoretical problem. Examples illustrate, however, that "the mental state of acquaintance" does not appear to be synonymous with the reductive referents of the concept I.

By many sleeping dream states or other states induced by drugs, hypnosis, meditation or traumata, experience by acquaintance appears to take place without being "brought under" the "I" (Howell, p. 19-20). The "I", on the other hand, can apparently experience itself as an agent behind actions (Epstein, p. 47-8). That violates the assumption that the mental state of acquaintance is purely passive, without agency.

Hypothetically, the reductive referents of the concept I and "the mental state of acquaintance" could nonetheless be intertwined phenomena. The "I" naturally appears to contain an "observer-observed duality" due to its self-reflective function. The observed part of the "I" may oft be its agent-part (Epstein, p. 47-8), yet may apparently also be its very observing part. The "observed observer" of each new second appears to require a "pure observer". The self-reflective process of the "I" may thus appear like a Russian Matryoshka doll with an endless amount of concentric layers. David Chalmers made a humorous note of a similar point (Chalmers, 1996, p. 230).

Sainsbury and Tye hold that we use a specific "I-concept" (presumably the concept I) to think about ourselves (Sainsbury & Tye, p. 144-5). Potentially, the observed part of the "I" (including the observed observer) corresponds to the concept I, while the *momentarily* observing instance is not part of it. Conceivably, the momentarily observing instance, when understood as detached from the concept I, is synonymous with the mental state of acquaintance. Thus, the reductive referent of both the concept PHENOMENAL CONSCIOUSNESS and the concept I could ultimately be identical.

A non-physicalist perspective

Concerning P-consciousness as such, the situation appears by first glance to be more complex than concerning a phenomenal character of discrete experiences such as those of sensory objects. Regarding colours, as example, the phenomenal side to a colour as neuro-functionally definable is "*the colour as it is phenomenally experienced*". Similarly, it appears that the phenomenal side to the mental state of acquaintance simply is "*the mental state of acquaintance as phenomenally experienced*". This cannot be thought of as being synonymous with P-consciousness. The same goes for the "I", since the "I" as phenomenally experienced is not the same as P-consciousness.

Still, the "I" and "the mental state of acquaintance" may display certain idiosyncratic properties worthy of further exploration. Firstly, P-consciousness and the mental state of acquaintance may conceivably "map onto" each other, meaning that experiential objects, features and qualities available to P-consciousness could be exclusively and exhaustively those accessible by acquaintance. This notion is also adaptable to Tye's view that P-consciousness simply *is* what is here called "the mental state of acquaintance" (Tye, p. 144-5). From a property dualistic perspective, furthermore, the mental state of acquaintance could be experienced as a state within which P-consciousness is solely present.

When it comes to the "I", next, P-consciousness could, within the present interactionist account, observe the observed portion of the "I" (including the observed observer) through the intimately experienced lens of the *momentarily* observing instance. The momentarily observing instance and the observed observer of the "I" could be experienced as synonymous (this will be further outlined). Next, given that the momentarily observing instance is synonymous with the mental state of acquaintance, the observed observer of the "I" could also be experienced as synonymous with the mental state of acquaintance.

The mental state of acquaintance could thus possess properties which allow a phenomenal illusion that it is synonymous with P-consciousness. We do not here talk about an illusion in the form of a cognitive belief. Rather, as according to the previously described principles, the person could simply be "blind" to the difference between phenomenal experience and experience by acquaintance. Here, specifically, he or she would be blind to the difference between P-consciousness as such and the mental state of acquaintance. If we accept a standard property dualistic rationale, this could in fact appear to be the natural end point of the current hypothesis.

Scrutinizing the interactionist hypothesis

Hypothetically, the brain activity which constitute the reductive referents of "the mental state of acquaintance" could become accentuated by an interactionist influence. This could cause LINS-activations in the same manner as assumed if, as example, the reductive referents of an experiential object like "vision of the sky" would be accentuated.

Since all experiences must be intimately linked to the presence of an experiencing subject, the interactionist accentuation of discrete experiential objects and of the mental state of acquaintance could potentially always be two sides of the very same event. If so, *all* interactionist influence could be thought to influence phenomenal judgements about both the present experiential objects as well as P-consciousness. We may call that scenario "unitary accentuation of object and subject".

As previously suggested, the experience of the LINS-activation could give rise to the illusion that there is something transcendent to the experience of that which is being accentuated. Concerning discrete experiential objects, simultaneous and co-dependent a priori experiences of LINS-activations and of experiential objects accentuated by an interactionist influence was assumed. Further, that only the discrete experiential objects have positively identifiable referents, so that LINS-activations are not sensed as independent experiential phenomena.

When trying to picture a similar rationale for cases of an interactionist accentuation of the mental state of acquaintance, we encounter more difficulty. Tye thinks the "realization" that (a) mental state(s) must exist which is "having" experiences by acquaintance (the mental state of acquaintance) only takes place a posteriori (Tye, p. 144-5). If correct, we must here assume the same to be the case also for P-consciousness.

Initially, Tye's view fits the notion that the mental state of acquaintance may be phenomenally *felt* as synonymous with P-consciousness. This, exactly because there is nothing about it which can be experienced a priori. Rather, it would operate more like an intimate lens for P-consciousness' experience of other phenomena. However, by unitary accentuation of object and subject, any phenomenal judgements resulting from confusion of that being accentuated and the LINS-activation, would thus appear to concern discrete experiential objects to the exclusion of concerning the mental state of acquaintance.

By separate accentuation of object and subject, on the other hand, LINS-activation following accentuation of the mental state of acquaintance would be experienced a priori without any a priori experience of anything at all being accentuated. This could, by analogy, potentially explain false possession of the concept PHENOMENAL CONSCIOUSNESS. That is, one could imagine that any reference to the LINS-activation as "phenomenal consciousness" or as "the mental state of acquaintance" would be more or less analogical to the naming of indigo as blue in the previous example of colour blindness.

The latter account would imply, however, that since the person does not know the actual mental state of acquaintance a priori, he or she would have no clear conception of what it really means. This, furthermore, does not do justice to what was assumed previously in this essay. It was then assumed that first person, a priori knowledge by acquaintance is both predominantly reliable and decisive for understanding phenomenal judgements.

From a non-interactionist perspective, it appears feasible that the a posteriori realization that a mental state of acquaintance exists is facilitated by awareness about the existence of discrete experiences by acquaintance. This may largely explain temporal co-occurrence of experiences by acquaintance and beliefs about a corresponding mental state having those experiences. It could also explain how one may realize that that mental state is ontologically linked to the character of experiences by acquaintance.

The challenge with the latter alternative is to explain how the insight about the mental state of acquaintance may suffice for the formation and use of the concept I. A requirement must be that the actual mental state of acquaintance, having the experiences by acquaintance which facilitate the a posteriori insight about that same mental state, will experience the object of that insight as a "solid" reference to itself.

It appears unclear, at best, whether the above is feasible, given that any a posteriori insight refers to an abstract, merely inferred object. Next, I will loosely outline an account which, in the best case, could aid both non-interactionist perspectives as well as interactionism past the above identified problems. Specifically, I suggest that the mental state of acquaintance may be experienced a priori, despite of Tye's conviction to the contrary.

We may imagine the mental state of acquaintance as "emitting" neuronal signals in a manner analogical to how light signals were emitted by P-consciousness within the previous "mirror analogy". As result of (non-interactionist) interaction with discrete experiential objects defined as neuro-functional patterns of brain activity, those neuronal signals could be altered, thereby coming to register and represent present experiential objects.

The mental state of acquaintance might have the capacity of possessing an underlying, intrinsic focus. This focus must not be understood as a reflectively conscious experience. By mere analogy to the visual mode of experience, the suggested focus might conceivably be compared to a direct experience of the sum total of brightness as was light an independent experiential phenomenon. That is, independently of the capacity of light, as medium, to define discrete experiential objects through colours and shades.

Since we talk about an intrinsic self-experience of the mental state of acquaintance, we must not assume that P-consciousness experiences the mental state of acquaintance a priori. This would have contradicted what was previously assumed. We must only assume that P-consciousness may experience, like may also the mental state of acquaintance, the a priori self-experience of the latter as some sort of brute quality.

A priori self-experience as here suggested should not be taken note of by the brain as something mysterious. Thus, the present account is fully adaptable to substance monist views like physicalism and versions of property dualism. Given interactionism, however, the neuro-functional manifestation of the a priori self-experience of the mental state of acquaintance may get accentuated by an interactionist influence. Further, the purported general confusion between any experiential object accentuated by an interactionist influence and the LINS-activation may then lead to the experience that there is a transcendent quality to the mental state of acquaintance. This, according to the exact same principles as outlined concerning discrete experiential objects.

Lastly, an "object" experienced a priori, such as here the mental state of acquaintance, appears sufficiently "solid" for the formation and use of the concept I. The reflective awareness about the self-experience held by the mental state of acquaintance could bring the mental state of acquaintance under observation and the concept I. Thereby, we might potentially also better understand how the momentarily observing instance and the observed observer of the "I" may be experienced as identical. Further, reflective awareness may logically only occur with a minimal delay relative to any referent, a priori experience. Since the a priori experience of the mental state of acquaintance may take place anew each second, we might thereby also understand the mentioned "Matryoshka-doll effect".

Conclusion

Initially in this essay, some problems were mentioned, the coherent solutions to which are instrumental for the notion that discussions of interactionism are warranted at all from a scientific perspective. The problems include those of causal closure of the physical universe, neuro-functional manifestation of potentially quantum level interactionist influences and the notion that causality requires functionality. Lastly, whether from a psychological perspective, thoughts and feelings could coherently be assumed transferred from a non-physical sphere to the physical brain.

In the central part of the essay, comparison was made to Michael Tye's physicalist perspective. I accepted Tye's division between knowledge by description and knowledge by acquaintance and that this represents an "explanatory gap". Distinctly, I suggested the existence of a "double" explanatory gap. I presumed a "phenomenal side" to all experience by acquaintance. This allows, also from the perspective of a non-physicalist hypothesis, that all discrete phenomenal experience is defined by physical brain processes. In isolation, this reflects a variance of a property dualistic rationale.

Because of theoretical complexity and space constraints, many aspects of the distinctly interactionist parts of the essay's hypothesis will remain treated exclusively in the main text. What should be mentioned, however, is that this essay purports that interactionist influences from P-consciousness upon the physical brain are reflex-like, content-less and subconscious. No discrete experiential objects are assumed "transferred" to the brain from a pre-existing state within a non-physical sphere. Importantly, however, the interactionist influence is still assumed to effect a process through which a distinct neuro-functional consequence, the so called "LINS-activation", may be experienced by the physical brain.

Two further features of the present hypothesis are worth emphasizing. It was assumed that "irrational convictions" about any transcendent or non-physical character to phenomenal experience may follow an interactionist influence. The present hypothesis necessitates the allowance of error, so that irrational convictions sometimes may be "just" irrational. Simultaneously, it accounts for the idea that the mentioned irrational convictions about phenomenal experience typically may result from first person experience of the neuro-functional consequence of the interactionist influence in the physical brain. Hence, the present hypothesis grants a very significant, albeit not necessarily all-decisive role to first person knowledge when it comes to explaining phenomenal judgements.

Lastly, the present hypothesis accounts not only for phenomenal judgements about the phenomenal character of discrete experiential objects such as thoughts or the sensory representation of physical objects, but also for ones about P-consciousness as such. Tye achieves the same by assuming that a reductively definable mental state (or states) is synonymous with P-consciousness. He thereby disputes the known doctrine that P-consciousness is functionally indefinable. The present interactionist hypothesis, on the other hand, accounts for beliefs that one may lingually refer to P-consciousness while also accepting its functional indefinability.

References

Arbeitsgemeinschaft für Methodik und Dokumentation in der Psychiatrie (2007) *Das AMDP-System. Manual zur Dokumentation Psychiatrischer Befunde* (8th Ed), Göttingen, Germany: Hogrefe.

Baker, M. C. & Goetz, S. (eds.) (2011) *The Soul Hypothesis. Investigations into the Existence of the Soul,* New York: Continuum.

Chalmers, D. J. (1996) *The Conscious Mind. In Search of a Fundamental Theory*, Oxford: Oxford University Press.

Chalmers, D. J. (2010) *The Character of Consciousness*, Oxford: Oxford University Press.

Eccles, J. C. (1994) *How the Self Controls Its Brain*, Berlin, Germany: Springer-Verlag.

Epstein, M. (2007) *Psychotherapy Without the Self. A Buddhist Perspective*, New Haven, Connecticut: Yale University Press.

Fahey, T. (2008) *Edmund Husserl*, [Online], http://www.tonyfahey.com/2008/12/edmund-husserl.html [03 March 2013].

Freeman, A. (ed.) (2006) *Consciousness and its Place in Nature. Does Physicalism entail Panpsychism?*, Exeter, UK: Imprint Academic.

Haisch, B. (2006) *The God Theory. Universes, Zero-Point Fields, and What's Behind It All*, San Francisco: Weiser Books.

Hirsch, S. R. & Weinberger, D. (eds.) (2003) *Schizophrenia Part Two Biological Aspects*, Oxford: Blackwell Publishing.

Howell, E. H. (2005) *The Dissociative Mind*, New York: Routledge.

Klinke, R. & Silbernagl, S. (1996) *Lehrbuch der Physiologie* (2d Ed), Stuttgart, Germany: Thieme.

Losee, J. (2011) *Theories of Causality*, New Brunswick, New Jersey: Transaction Publishers.

Ogden, T. H. (1986) *The Matrix of the Mind. Object Relations and the Psychoanalytic Dialogue*, Lanham, Maryland: Rowman and Littlefield, 2004.

Papineau, D. & Selina, H. (2000) *Introducing Consciousness*, New York: Totem Books.

Popper, K. & Eccles, J. C. (1977) *The Self and Its Brain. An Argument for Interactionism*, Abingdon, UK: Routledge, 2003.

Sainsbury, R. M. & Tye, M. (2012) *Seven Puzzles of Thought and How to Solve Them: An Originalist Theory of Concept*, Oxford: Oxford University Press.

Sartory, L. (1996) *Understanding Relativity. A Simplified Approach to Einstein's Theories*, Berkeley: University of California Press.

Smolin, L. (2001) *Three Roads to Quantum Gravity*, New York: Basic Books.

Stoljar, D. (2010) *Physicalism. New Problems of Philosophy*, Abingdon, UK: Routlegde.

Tye, M. (2009) *Consciousness Revisited. Materialism without Phenomenal Concepts*, Cambridge, Massachusetts: The MIT Press.

Review Article

Pineal Gland, DMT & Altered State of Consciousness

Iona Miller*

ABSTRACT

Endogenous and synthetic DMT and their relation to pineal function have been the subject matter of numerous popular and scientific articles, crossing many disciplines. Correlations for visionary and psi-conducive states have been widely suggested. Numerous spiritual technologies have been retrieved from traditional cultures and new technologies based in frequencies and resonance have been concocted and sold to the public. Yet, in 2010, even the leading researcher Rick Strassman says that "we don't know if DMT does appear in the pineal…[or, if] endogenous DMT activity increases during particular non-drug altered states, such as dreams and near-death experiences." If it were, it would help explain the psychedelic characteristics of those altered states."

This strongly suggests that new assays for low levels of endogenous DMT, 5-MeO-DMT, bufotenine, and metabolites in different tissues would be very useful. Such experiments were conducted in 2012 with positive results. Nevertheless, a review of the history and speculation on the psychoactive compounds remains of great interest to both researchers and the general public. What do these things mean? The jury remains out, and we are wise to remember that theories remain just that. Baselines need to be established for normal waking consciousness, and comparisons made for a variety of states of consciousness. But perhaps the greatest result of such research is new understanding of what it means to be fully human.

Key Words: pineal gland, DMT, dimethyltryptamine, N,N,-dimethyltryptamine, 5-methoxy-tryptamine, altered state of consciousness, 5-MeO DMT, "telepathine", Pinoline, 6-methoxy-tetra-dydro-beta carboline, 6-MeO-THBC, hallucination, darkroom retreats, soma pinoline, shamanic journeys.

I am created by Divine Light. I am sustained by Divine Light. I am protected by Divine Light. I am surrounded by Divine Light. I am ever growing into Divine Light. Swami S. Radha in Realities of the Dreaming Mind (1990).

Introduction

What do imaginative children, passionate lovers, dreamers, psychonauts, telepaths, bliss-bunnies, UFO abductees, shamans and neo-shamans, birthing mothers and babies, near-death experiencers, and schizophrenics have in common? The same thing Tibetan, Taoist and Kabbalistic masters, meditators, mystics and religious prophets share.

Popular theories contend that their brains may be flooded with natural psychedelic pineal

*Correspondence: Iona Miller, Independent Researcher. Email: iona_m@yahoo.com Note: This work was completed in 2006 and updated in 2013.

secretions that tenaciously cling to their synaptic junctions, electrifying their whole being with multisensory virtual stimuli, experiential beliefs and delusions about the nature of reality.

Can the resonant buzzing of our own bioelectronic Third Eyes awaken latent metaprogram circuits, leaving us virtually blinded by compelling mind-altering hallucinations? In extraordinary circumstances, chemical brainstorms of scintillating multidimensional visionary Light are generated in our own craniums producing demons, divinities, aliens, oceanic ecstasies, death/rebirth, and time distortion. According to some, this hyperdelic retinal circus is your brain on the natural high, DMT.

Whether or not DMT can be proven as the source, there is a long history of unusual phenomenology contained in the world's great literature and in traditional lore. This spiritual supercharge, revealed in all wisdom writings, has been sought by cultures from Egypt and Asia to the Americas, from Druids to Tibetan lamas, and now in modern science.

Psychedelic alchemist, Sasha Shulgin claims, “DMT is everywhere,” in nature. Seekers have ceremonially emptied or exhausted their mindbodies, using plant or animal products, spiritual technologies, sensory deprivation, social isolation, fasting and ordeals, rhythmic chanting, drumming, electronic resonance, even sex, to hasten and drive the neurological illumination. (Miller, 2006)

DMT was ‘discovered’ by Richard Manske (by synthesis) in a lab in 1936, and later found in South American shamanic plants in the mid-1940’s. But, its psychoactive properties were not recognized until 1956 by the Hungarian chemist ad psychiatrist Stephen Szára. With his colleagues, he characterized the biochemistry of the first three psychedelic cogeners of tryptamine: dimethyl-, diethyl-, and dipropyl-tryptamine (DMT, DET, and DPT), describing their pharmacological dynamics and effects.

Mathematician Ralph Abraham (2006) reports scientific inspiration from synthetic DMT:

“*At one time, around 1969, we used large doses of DMT, and this period was crucially important to the whole evolution of my mathematical understanding of consciousness, based on geometry, topology, nonlinear dynamics and the theory of vibrating waves. For in these experiments, although lasting only a few minutes, the reciprocal processes of vibrations producing forms and forms producing vibrations were clearly perceived in abstract visual fields.*”

Professor Steven Barker has taken DMT research into the philosophical level, as well, considering questions of spirituality and the nature of belief in God. He focuses not only on the pharmacology and hallucinations, but reports of spiritual effects and religious experience. Such preparations give the impression of "being in touch with the universe". Drugs such as ayahuasca have inspired a robust tourist trade and an explosion of shamanic experts of varying credibility.

Barker reports, "[These compounds] cause euphoria, tunnels of light, they see fantastic beings — deities, relatives — you can't explain it. Those phenomena ... we know these compounds can do those things." He admits such experiences can be kindled without ingesting drugs. He calls such spontaneous phenomena the biological and molecular basis for religious faith. This is how "God" emerges from our bodies. Speculations have given way to the emerging science of transcendental experiences -- neurotheology, the anatomy of spiritual life.

This Is Your Brain on Youth

We are hardwired to seek pleasure, to seek ecstasy. As children we expressed our authentic, core selves – that which can neither be taught nor learned. But the whole "self-help" and "new age" movements are based on trying to get back that golden state of innocence, childlike wonder and awe with spiritual connection.

When religion suggests we "become as little children" again, who could imagine this implies a shamanic return to the womb and the natural psychedelic state of an uncalcified pineal gland? In early childhood, we are perpetually immersed in cascades of trance-inducing theta rhythms of the brain, with the feel-good chemical brew it creates for metaprogramming.

Until roughly age 8, we can't really distinguish between fantasy and reality, due to our own natural hallucinogen, DMT, (dimethyltryptamine). DMT molecules are similar to serotonin and target the same receptors. Meditation has been suggested as a means of preserving youthful appearance and mental flexibility. Dr. Rick Strassman and others claim this spiritual technology encourages production and release of natural DMT (Soma Pinoline) in the mindbody throughout the lifespan.

DMT is implicated in the wild imaginings of our nightly dreams, near-death phenomena (NDEs), alien abduction experiences, and dream yoga. It is also a plausible source of visionary phenomena in therapy, such as unusual psychophysical states attained in waking dreams, shamanic or psychotherapeutic journeys. Synthetic and botanical DMT crosses the blood-brain barrier and bonds to the same synaptic sites as serotonin.

"Each night in dreams we experience an essentially psychedelic state. The principal difference between dreams and hallucinations is the way the stages of wakefulness are organized, with the suppression of REM sleep and the intrusion of PGO waves in the arousal (waking) stage and in NREM (or slow) sleep.

"The stages include: waking (arousal) stage, stage of PGO waves, hallucination stage, sleep stage and hallucinatory manifestations. The waking dream eliminates "residues" stirred up by the PGO wave pattern in the absence of REM sleep. These visions resemble those at the approach of death, or what are called near death experiences (NDEs). In another context, they are perceived as visions. They include the characteristics of two phases of NDEs (Sabom, 1982):

Autoscopic phase includes: *1) subjective feeling of being dead; 2) peace and well-being; 3) disembodiment; 4) visions of material objects and events.*

The Transcendental phase includes *5) tunnel or dark zone; 6) evaluation of one's past life; 7) light; 8) access to a transcendental world, entering in light; 9) encounter with other beings; 10) return to life.*

The Path of Light: Primordial In-Sight

The pineal sits, well protected in the deep recesses of the brain, bathed in cerebrospinal fluid by the ventricles, the fluid-filled cavities of the brain that feed it and remove waste. The ventricles also function as resonant capacitors, sensitive to certain frequencies. There are many scientific examples of psychedelic body fluids and metabolites (Collected Abstracts).

The pineal gland emits its secretions to the strategically surrounding emotional, visual and auditory brain centers. It helps regulate body temperature and skin coloration. It secretes the sleep hormone melatonin, which is also implicated in DNA repair (Santoro) and epigenetic regulation (Korkmaz). Epigenetic, inherited genetic modifications are known to be involved in disease.

Generally, after the more imaginative period of childhood, the pineal calcifies and diminishes at the onset of puberty's sex hormones, around age 12. The pineal is the only unpaired gland in the brain. Curiously, this solitary gland is light sensitive and actually has a lens, cornea, and retina. Is there a retinal circus of biophotons deep in the brain, which only the Third Eye sees or even creates: the Light of Wisdom?

This "Third Eye," in the center of the brain, is implicated in the production of endogenous or natural DMT, dubbed the Spirit Molecule in popular literature. The pineal synthesizes natural hallucinogens in response to certain psychophysical states, and raises serotonin levels in the brain. There is a functional decline in the gland with advancing age.

This master gland is responsible for the internal perception of Light, the raising of Kundalini the serpent power, and for awakening inner sight or in-sight. The key to a successful meditation is the withdrawal of the sensory currents to the eye focus or the third eye. Once there, the gaze focuses on the middle of whatever appears without any distractions or intrusive thoughts, ideally immersed in the unconditioned field.

The groundbreaking work of Dr. Rick Strassman (2001) focuses on the role natural body chemistry plays in creating spiritual life. He calls DMT the Spirit Molecule; an endogenous hallucinogen, which he boldly asserts, is an active agent in a variety of altered states including mystical experience. To explore his theory, Strassman conducted extensive testing, injecting volunteers with the powerful psychedelic, synthetic DMT (N,N-dimethyltryptamine; N,N-DMT).

DMT is so powerful it is physically immobilizing, and produces a flood of unexpected and overwhelming visual and emotional imagery. Taking it is like an instantaneous LSD peak. DMT crosses the usually impenetrable blood-brain-barrier, suggesting its fundamental role in consciousness. But, concluding his 5-year studies early, Strassman admitted despite their growth potential, there were no viable therapeutic or neurological applications. He does *NOT* recommend recreational use.

DMT production is stimulated, in the extraordinary conditions of birth, sexual ecstasy, childbirth, extreme physical stress, near-death, and death, as well as meditation. Pineal DMT also plays a significant role in dream consciousness. This chemical messenger links body and spirit. Pineal activation awakens normally latent neural pathways.

"All spiritual disciplines describe quite psychedelic accounts of the transformative experiences, whose attainment motivate their practice. Blinding white light, encounters with demonic and angelic entities, ecstatic emotions, timelessness, heavenly sounds, feelings of having died and being reborn, contacting a powerful and loving presence underlying all of reality--these experiences cut across all denominations. They also are characteristic of a fully psychedelic DMT experience. How might meditation evoke the pineal DMT experience?"

"Meditative techniques using sound, sight, or the mind may generate particular wave patterns whose fields induce resonance in the brain. Millennia of human trial and error have determined that certain "sacred" words, visual images, and mental exercises exert uniquely desired effects. Such effects may occur because of the specific fields they generate within the brain. These fields cause multiple systems to vibrate and pulse at certain frequencies. We can feel our minds and bodies resonate with these spiritual exercises. Of course, the pineal gland also is buzzing at these same frequencies. . .The pineal begins to "vibrate" at frequencies that weaken its multiple barriers to DMT formation: the pineal cellular shield, enzyme levels, and quantities of anti-DMT. The end result is a psychedelic surge of the pineal spirit molecule, resulting in the subjective states of mystical consciousness." (Strassman, 2001).

Natural hallucinogens may belong to the tryptamine or beta-carboline family of compounds. One compound (6-methoxy-1,2,3,4-tetra-hydro-beta-carboline) has been implicated in rapid eye movement sleep (REM). It is concentrated in the retinae of mammals, which may be related to its visual effects. There are several ways in which either psychoactive tryptamines and/or beta-carbolines may be produced within the central nervous system and pineal from precursors and enzymes that are known to exist in human beings. In addition, nerve fibers leave the pineal and make synaptic connections with other brain sites through traditional nerve-to-nerve connections, not just through endocrine secretions.

Third Eye Blind

Serotonin or tryptamine levels are higher in the pineal than any other organ in the brain. 5-methoxy-tryptamine is a precursor with hallucinogenic properties, which has a high affinity for the serotonin type-3 receptor. Gucchait (1976) has demonstrated that the human pineal contains an enzyme capable of synthesizing both DMT and bufotenine-like chemistry. These compounds are prime candidates for endogenous "schizotoxins," and their production may be related to stress and/or trauma, that correlate with schizophrenia.

Strassman notes that both the embryological rudiments of the pineal gland and the differentiated gonads of both male and female appear at 49 days. Melatonin is a timekeeper for gonadal maturation, so the pineal is implicated again. He suggests this rein-effect may be the root of the tension between sexual and spiritual energies, yang and yin. The pineal gland is a source of both psychedelic compounds and the regulates the gonads, sources of spiritual and generative immortality.

Stress-related hormones cue the pineal activation to activate normally latent synthetic pathways, creating tryptamine and/or beta-carboline hallucinogens. When we face stress or potential death, or in meditative reveries, we "tune back" into the most well developed motif of such experiences--the birth experience. Perinatal themes and memories re-emerge (Grof).

Those with Cesarean deliveries report greater difficulty in attaining transcendent states of breakthrough and release during drug-induced states. Maybe less fetal (or maternal) hallucinogens were released at the time of birth. They may not, according to Strassman, have a strong enough "*template of experience*" to fall back on, to let go without fear of total annihilation, because lesser amounts of pineal hallucinogens were produced during their births.

Through meditation, the pineal may be modulated to elicit a finely tuned standing wave through resonance effects. It creates the induction of a dynamic, yet unmoving, quality of experience. Such harmonization resynchronizes both hemispheres of the brain. It recalibrates the whole organism.

Dysynchrony is associated with a variety of disorders. Such a standing wave in consciousness can induce resonance in the pineal using electric, magnetic or sound energy, and may result in a chain of synergetic activity resulting in the production and release of hallucinogenic compounds. Thus, the pineal can be likened to an attractor, or "lightning rod" of consciousness. It generates an illuminative laser beam that pervades the energy body.

A Walk on the Wild Side

McGillion (2002) reports that ancient cosmobiologists noticed effects of the planets were mediated by the pineal gland. Physician-astrologers in Greece and Europe assumed a correlation between events in the heavens and those on earth that was relevant to both health and disease.

The sun, moon, and planets studied by the early health practitioners can, and do, affect us. A simple example is being born in the day or night can make us a "morning person" or "night owl," by setting the circadian body-clock of pineal sensitivity and melatonin production.

Humans are among the organisms in which light sets off chain reactions of enzymatic events in the pineal gland that regulate circadian rhythms. It modulates photoreceptive circadian oscillators, sets the pacemaker, and governs longer cycles. Seasonality and geomagnetism at birth can affect long term development. In this way, it influences our relationship with people, animals and the earth. It keeps us in synch with life.

Physicist Cliff Pickover argues that, "DMT in the pineal glands of Biblical prophets gave God to humanity and let ordinary humans perceive parallel universes." "Our brain is a filter, and the use of DMT is like slipping on infrared goggles, allowing us to perceive a valid reality that is inches away and all around us."

He suggests, perhaps our ancestors produced more DMT, leading to extraordinary spiritual visions. "Maybe this is why the ancients seemed so in touch with God and with miracles and visions. Maybe Moses, Mohammad, and Jesus had a greater rate of pineal DMT production than most." Pickover blames artificial light for a reduction in our DMT production rate. Or perhaps more likely, as most ancient cultures, they simply supercharged themselves with shamanic herbs.

Some claim the "burning bush" was *Cannabis sativa*, Assyrian Rue (*Pegunam harmala;* Zoroaster's Hoama, Asena) or the North African Acacia tree, and that Moses either smoked the leaves or got high downwind in the DMT-containing smoke. Graves, (1948, 264) claimed the Acacia *Sant*, a host tree of the mistletoe-like *loranthus*, was the 'burning bush' and source of manna.

If this oracular tree of Canaan contained tryptamines, as many species do, Moses could have had access to DMT illumination. It is still a practice to burn botanicals inside a tent to imbibe their smoke. Assyrian rue was the most sacred plant of Mohammad, who took the *Esphand* (Arabic/Persian name for the plant) before receiving the Koran from God. This holy *Esphand* was associated with the appearance of angels and casts out evil spirits, and was used to cure fever and malaria and provides the rich red dye of Persian carpets.

Rue was central to the Petra mystery rites and schools of alchemy. Their sacrament was a beverage of illumination and restoration, mixed with gold and other alchemical products. It was passed down from Zoroaster, who was also known as Chem the original Alchemist and CHEMist. Chem is an ancient name for Egypt. The grandson of Zoroaster, Nimrod or King En.Meru.dug, founded the Egyptian 2nd Dynasty. This plant of life became central in the Mysteries and healing schools of ancient Egypt, where Moses could easily have learned its powers. (Bosman)

The original Essenes, named after the plant, were headquartered in Heliopolis. *Asena*, the botanical Bush of Life, is an acronym, ASNA, of Aset (Isis) Sutekh (Set) Nebtet (Nephtys) and Auser (Osiris). It embodied the female form of the One God ATON, who correlates with the Greek goddess of wisdom, Athena.

Shamanic Bedouins still make the Egyptian eucharistic Bread of Light using the Asena/Hoama bush, the North African Acacia tree, and ground meteorite. They still shape it the form of the Eye of Ra, a circle with a central hole. The tradition was passed down to Christian Gnostics in Abydos, Egypt where the bush eventually was used symbolically to sprinkle holy water. Leonardo da Vinci and Michaelangelo reportedly used it for visionary inspiration.

In modern times, Pickover has suggested spontaneous DMT experiences as a possible source of Whitley Streiber's *Communion* aliens: "…we know that DMT can often produce visions of cartoon-like aliens." Others (Meyer, 1993) report those using DMT often claim communication with stick-like, insect-like or elf-like beings and discarnate entities. Psychonaut, Terence McKenna used synthetic DMT (N,Ndimetyltryptamine) to intentionally contact "machine elves" and explore parallel worlds: "What is driving religious feeling today is a wish for contact with this other universe."

Arguing for their radical "otherness," McKenna emphasized that the hyperspace aliens seen while using DMT present themselves "with information that is not drawn from the personal history of the individual." But smoking this synthetic DMT for 'hyperspace experience' rarely yields Clear Light experiences.

The Harma alkaloid "Harmine" is also known as Telepathine and Banisterine. It is a naturally occurring beta-carboline that is structurally related to harmaline. They stimulate the CNS by inhibiting the metabolism of serotonin and other monoamines. Telepathine, is an MAO inhibitor, which parallels the function of Pinoline (a natural MAO Inhibitor) naturally produced by the Pineal Gland. The combination of the Pineal secreted DMT (Dimethyltriptamine) and the MAO Inhibitor, Pinoline (Methoxytetrahydrobetacarboline, MeOTHBC) may be responsible for naturally occurring psychic experiences.

Harmine and harmaline are found in Syrian Rue (3-7% harma alkaloid), and ayahuasca brews made with DMT sources, bark and leaves of "Pychotriaviridis" or *Banisteriopsis caapi* vine. Ayahuasca is the South American sacrament of the Church of Santo Daime and Uniao de Vegetal (UDV). In this setting, the churches condition the "spiritual" expectations, experiences, ethics and type of information "received" in the altered state. The visionary state is considered to be the essence of the shamanic complex.

Shamanic vision differs from hallucination in volition, form and content of thoughts, clarity of heightened awareness, perception and contextualization. Practitioners have claimed it is for "analysis". There is no primary delusional experience. The distinction between self and non-self is blurred; the notion of causality is affected. Its chemistry probably works as follows:

The primary function of harmala alkaloids in ayahuasca is to allow for the oral activity of DMT by inhibition of MAO-A, and further permits accumulation of 5-HT and other neurotransmitters. On their own harmala alkaloids have only weak psychoactive effects (Callaway, 1994) but Kim et al (1997) found that the harmala alkaloids, which occur in ayahuasca, were the most effective inhibitors of purified MAO-A. (Callaway, 1994)

The psychedelic effects of ayahuasca probably manifest primarily through the serotonergic effects of DMT on the CNS and through increased levels of unmetabolised biogenic amines. Pinoline potentiates the activity of methylated tryptamines and this is the probable mechanism behind ayahuasca (Callaway, 1994) Investigation of long-term users of ayahuasca showed a statistically significant difference between control group and users with a higher binding density in blood platelets of 5-HT uptake sites in the ayahuasca drinkers. This indicates a modulatory role for pinoline (the endogenous equivalent of ayahuasca) in the CNS. An upregulation of the serotonergic system is exactly what current antidepressant medications attempt to do, i.e. increasing synaptic 5-HT by preventing its reuptake. (Callaway, 1994)

Phalaris grass expert, "Johnny Appleseed" brings the finesse of a professional Transpersonal Psychologist to his experiential shamanic teaching. Phalaris arundinacea and Phalaris aquatica, as well as Phalaris brachystachys,contain the psychoactive alkaloids N,N-DMT, 5-MeO-DMT, and 5-OH-DMT. He contends 5-MeO DMT awakens psychic centers by amplifying our telepathic ability to affect or be receptive to others' brainwaves through modulating chemistry, electromagnetic entrainment and standing feedback loops. Plants are mixed in brews with MAO Inhibitor containing plants (Banasteriopsis Caapi, Syrian Rue etc.), to produce entheogenic brews that mimic the DMT-Pinoline combination naturally produced by the pineal gland in the brain.

"*For the last ten years I have been doing healing and exploratory sessions with individuals and small groups. We have occasionally experienced the phenomena of telepathy, ESP, and interactions on an energetic level that produce healing in a number of modalities. I work only with a strain of Phalaris grass, an entheogen specifically bred to contain 5-MeO DMT. I use this in an oral preparation potentiated by MAOI from Syrian Rue. This is not a common mix, as most people smoke it for a blast, which is not conducive to this work. I feel all of the other materials mentioned, LSD, mushrooms, and DMT-based brews distort consciousness to some degree. Oral-based 5-MeO DMT from plant sources is something completely different, however. Traces of other alkaloids from the plant source produce a more enhanced experience than the pure chemical. A clear yet enhanced state is accessible easily and reliably with no delusional ideation, or visual distortions.*

It is simply like being fully awake. This is probably because we were born with, and until puberty had, a pineal gland that made 5-MeO DMT in quite substantial amounts, unlike the reports of only very trace amounts of endogenous DMT. Thus, we have the receptors and metabolic pathways to deal with this material in a non-distorting way." (Appleseed, private correspondence, 2001).

The pineal gland makes a neurohormone called melatonin, which is one of the key regulators of the circadian and seasonal biological rhythms. It also makes a mono-amine oxidase (MAO) inhibitor called pinoline (Methoxytetrahydrobetacarboline (MeOTHBC)) which acts on the GABA receptors and whose chemical structure is virtually identical with the harmala alkaloids. Serotonin (5 Hydroxytryptamine (5HT)) has frequently been implicated in certain aspects of psychoses.

Pinoline is a neuromodulator, which prevents, amongst other effects, the breakdown of serotonin. This results in an accumulation of physiologically active amines including dimethyltryptamine (DMT) within the neuronal synapses, which may lead to hallucinations, depression or mania depending on the amines being affected (Strassman, 1990).

Ananda M. Bosman emphasizes the crucial role of endogenous DMT and sacramental DMT from Syrian Rue and other potentiating botanicals. The ancients called it *Hoama* in Persian (*Avesta Veda*), *Soma* in Sanskrit (*Rg Veda*), the Egyptian *Essene*, the Sumerian Tree of Life, Mohammed's *Esphand*, the burning bush *Asena* of Moses, the Gnostic *Besa*, the Etruscan *Phallaris arundanacia*, and the *Rue* of alchemy.

Syrian Rue was revered because Melatonin's active metabolite Pinoline is oneirogenic and antidepressant, increasing Serotonin turnover. Lack of Pinoline disturbs our circadian rhythms and creates depression. Pinoline has been conclusively demonstrated to have no function in schizophrenia, since test subjects healthy and otherwise, had the same levels of Pinoline. (Bosman)

The pineal is a superconducting resonator. Ananda claims it potentiates DNA as a multidimensional transducer of holographic projection, through hadron toroids, and is implicated in staying youthful. 5meoDMT and DMT act on the T-RNA messengers, which carry out the protein synthesis for the DNA, or the rebuilding of our body image and organs.

Melatonin is exclusively made in the pineal gland, comprised of the same Tryptophane base materials as Pinoline. Melatonin induces mitosis. It does this, by sending a small electrical signal up the double helix of the DNA, which instigates an 8 Hz proton signal that enables the hydrogen bonds to the stair steps, to zip open, and the DNA can replicate. The human Pineal gland not only produces the neuro-hormone ,elatonin, one of the body's most potent antioxidents, but the revolutionary Pinoline, 6-methoxy-tetra-dydro-beta carboline, or 6-MeO-THBC.

Pinoline is superior to Melatonin in aiding DNA replication. Pinoline can make superconductive elements within the body. It encourages cell division by resonating with the very pulse of life - 8 cycles per second - the pulse DNA uses to replicate and the primary Schumann Resonance. Andrea Puharich measured this 8 Hz resonance in healers in the late 1970s.

Ananda implicates DMT in the hyperdimensional geometry or architecture operating in DNA through hadronic mechanics, a model of the 8 Hz, or universal phase-conjugational force, that is also the most coherent Nuclear Magnetic Resonance and the DNA replication frequency. He relates this to the sacred geometry of the Merkabah, Flower of Life, Sri Yantra, Diamond Body, and Vector Equilibrium Matrix.

The living DNA in our bodies operates in the hyperdimensions of wave-genetics. The entire body holographic message is present in the single DNA molecule, in order to be capable of reproducing the entire whole. The local 8 Hz field component is a standard tetrahedron interlocked with a second tetrahedron representing the counter-rotary field that it is phase-conjugating with, which together comprise a "stellated cube."

Ananda has also investigated Dark Room techniques for stimulating the pineal with chi master, Mantak Chia. Ananda developed, researched, and has taught the Dark Room technology for endogenous Pineal Soma and DMT production (Endohuasca), since 1992, upgrading his technique with Master Chia in 2000. Isolated from external light, the third eye (pineal gland) overflows with certain neurotransmitters that awaken the higher brain, the ability to imprint the brain, reprogramming itself for an "instant experience of Being."

The retreat claims to reopen the source code of embryogenesis. 5-MEO-DMT activates the whole spine, the whole tree of life (Djedi, the staff of Hermes, the Caduceus of the spine) becomes active to be reprogrammed. This is the accessing and awakening of the tree of life, the kundalini, which is a readout of the DNA. DNA itself is a minute tree of life.[See "Pitch Black", below]

In *The Unity Keys of Emmanuel* and *Somajetics*, Ananda says, *"[By the DMT translation of the Sound of Silence of the Word into the Soul Computer Virtual Reality Interface, the parallel quantum bodies are thus accessed by the NMDA inhibition, which is electrical anesthesia, engaged by the heart ecstasis of 8 hz, to the 1000 hz petaled lotus.* [This] *is the learned means of NMDA inhibition, engaged through the cave and dark room retreats of inner re-engagement. Thus the chemical soul's crystal laser transducers and interdimensional door keys, are in Soma Harmaline-Pinoline-Harmine, DMT and 5-MeO-DMT, and by the NMDA inhibitors through ecstasis."*

"A high spin state within the DNA water molecule harnessed by Pinoline/Soma intercalating with the DNA (a molecule that has a stable 8hz NMR proton-proton spin-spin coupling [thus hadron pi-meson interplay), together with N-Methl-D-Aspartate-Inhibitors enables the electron states to move into a nulling, and electron freeze within an entire cell, enabling only 8 hz fields to pass through, changing the charge of the cell, so that the superconductivity harnessed by the pinoline DNA intercalation (with sonic interaction of the vocalized DNA electron spin resonance tones) — enables the hadronic force to be able to operate within the macro region of an entire cell (and intercellularly, by extension)."

Furthermore, noradrenaline plays a significant role in the Pineal gland, when there is sufficient Pinoline saturation in the brain. It releases a serotonin site, enabling another serotonin site on the pineal gland to produce the potent visionary Dimethyltryhptamine (DMT), neurotransmitter.

Dr. James Callaway detected this molecule in the spinal serum of people who were dying, or were having an "Out of body experience (OOBE)", or who were lucid dreaming. It is Pinoline that enables the threshold levels of DMT to become active in the brain, but it requires an adrenaline burst. DMT with Pinoline increases brain activation, and with its cousin the 5-Methoxy-DMT, has been shown to activate the brain by as much as 40%, compared to our 10% maximum potential at present. This is a frightening prospect for the uninitiated, due to the absolutely overwhelming nature of DMT.

Youthenize Yourself

Hollywood trainer to the stars, Barry Hostetler, P.I. advocates "Youthenizing", claiming his 35 years of bodybuilding and meditating allows him to "kick out the DMT." We can control our psychobiology by controlling our mental state and vice versa, especially the reactive "dragon brain" or "reptilian brain". Paramahansa Yogananda taught, "If you are in a dark room, don't beat at the darkness with a stick, but rather try to turn on the light!"

Under stress we release toxic catabolites into our system, which undermine the immune system and age us faster. Without exercise (aerobic, core, strength) poisons accumulate, making the body toxic and mind frustrated and agitated. When we are calm and balanced, body chemistry is nontoxic and immune function improves.

Meditation is an alpha brain wave entrainment technique, which synchronizes the two brain hemispheres into 8 Hz. Closing the eyes, stops Melatonin flow leakage to the body, and makes it saturate the neocortex, increasing concentrations of Meltaonin and Pinoline in both brain and body. Meditation, several times a day, is an essential health exercise, an energizer, and tool for mental integration of daily activities.

Pinoline and related beta carbolines are not only produced in the brain, but in the adrenal glands themselves, where these hormones undergo their transformation to the hormones of life. HeartMath Institute demonstrated that minutes of compassion in the cardio-rhythm, which induces 8 Hz in the brain, brought DHEA up to youthful levels.

Twenty minutes of compassionate love, through meditative breathing, and whole body 8 Hz entrainment is the ultimate hormone precursor anti-aging pill. Not only does the pineal gland produce more Melatonin and Pinoline, which instigate 8 Hz ELF waves throughout the body, but these neurohormones signal the pituitary to release the life hormone Somatropin, which signals the adrenal glands to instigate cholesterol to convert to Pregnelenone then DHEA.

The extra Pinoline and other beta carboline levels that result, aid the body cells to replicate, and neutralize microorganisms, parasites, fungoids, and bacterias, and related harmful invaders. Melatonin and Pinoline are also antioxidents. Meditation is a rest break, an exercise session, an integration session, an energizer, and a body tuner, promoting antioxidant and antidepressant production. This makes meditation valuable for stress-management and simple self-care.

We can return intentionally to more youthful states by doing emotional exercises and visualizations, which stimulate the body chemistry of our glory days. The body remembers and mimics those chemical states, producing youthful hormones and more flexible mental and physical states, improving overall balance and disposition. When "Youthenizing" yourself, it is helpful to use a photo from under age 7, a time you felt at your peak, or your happiest, or other 'good chemistry' times.

Kinesiology demonstrates that the mind "thinks" with the body itself. Mindbody is the subtle mechanism behind the disease process. The chemistry you generate with moods and states in your body is crucial to your health and well being. First toxic states of mind affect the energy body, then the physical body. Subjective and objective experience are hidden determinates of behavior. Embodied as corporeal memory, the body *is* your memory and subconscious. Self-regulation can modulate this process.

The body is an island of energy/matter and emotions with waves of feelings crashing onto its shores. Body-consciousness can either hide or reveal spirit, depending on how we direct our attention toward our ego, stress (including spiritual distress) and relief. The body and mind can be reunited in a congruent, healthy lifestyle by acting on what you know.

Pitch Black

Absolute darkness has an initiatory quality – the metaphor of moving from the darkness of ignorance into the illuminative Light. But it is more than a metaphor. All wisdom traditions have used sensory deprivation and darkness, (such as caves, tunnels, catacombs or special chambers), as a shamanic mind-altering force. Disorientation outside facilitates internal focus and connection. Dark Room (DR) technology is for core reprogramming, restimulating the hardened pineal which begins calcifying around age 12. Sound becomes light. Chanting and drumming amplify the effects, which culminate in a rebirth of the spirit when one enters the point of Light or primordial Luminosity.

Author, Robert Newman, (*Calm Healing*, 2006) advocates Medicine Light for a variety of conditions. He cites Tibetan Buddhist, Trungpa Rinpoche about a highly advanced, dangerous form of meditation practiced in utter darkness, known as a Bardo Retreat. They last up to seven weeks in a specially prepared Darkness Chamber, during which the whole *Tibetan Book of the Dead* is experienced and visions arise innately from the brain. The beneficent and wrathful "eyes of Buddha" become visually and interactively alive as the Bardo of Luminosity flashes on and off; visions are self-arising; and transcend ordinary perception. The 49-day cycle recapitulates embryogenesis up to the point when pineal and gender differentiation occurs. The Taoists, Egyptians, Druids and others had similar practices of external light isolation.

Taoist master, Mantak Chia of Thailand recommends sound and light isolation in a process he calls Darkroom Enlightenment. He sequesters participants in the dark for over a week to shock the pineal into critical arousal to stimulate production of natural DMT and break down the barriers to transcendence. The neurotransmitter 5-MeO-DMT is normally only active when we are in the womb and in the first months of our lives. It is reactivated in the darkroom.

Stages include the 'Melatonin state (Day 1-3; ego death), Pinoline State (Day 3-5; energy body and astral projection; lucid dreaming), 5-MeO-DMT (Day 6-8; telepathy, White Light), culminating in illuminative DMT (Day 9-12; Clear Light; Immortal Body). Participants leave through a tunnel, presumably a symbolic rebirth.

"*There is now enough 'Mono Amine Oxidase Inhibition' triggered by the pinoline, to allow the pineal gland's 'serotonin to melatonin cycle' to be intercepted by adrenaline and ephedrine activity and converted into a 'serotonin - DMT pathway'. When DMT levels reach more than 25mg, one's experience can become very visual. DMT is the visual third eye neurotransmitter. It enables the energy body and spirit to journey into hyperspace, beyond third dimensional realms of time and space.*" (Chia, 2006)

Reentry implies rebirth -- the self-organizing emergence of the new self. The seed of initiation is realized as the mature fruit of experience, which feeds and sustains us. The experience continues to be useful in our lives. Each healing journey is the death of something within us, which has kept us stuck or stultified. Healing facilitates our continuing evolution. The new self continues to emerge and the consequences of the journey become embodied in this new form for months and even years after the journey. New behaviors, feelings, attitudes, ideas, and wisdom follow. Thus, the circle of life continues unbroken.

Entheogen expert, K. Trout (2001, correspondence) is skeptical: "I'll ignore the shakiness of their biochemical presentation but to arrive at DMT from melatonin or a betacarboline starting point would be interesting if not fanciful under physiological conditions. It is much more likely that formation of endogenous DMT would be instead of melatonin rather than from it or into it (and the two paths mutually preclusive similar to what we see in plants as concerns DMT and 5-MeO-DMT synthesis even when the two co-occur). Now to go from melatonin to a 6-

methoxylated-betacarboline with psychoactivity on the other hand is not at all far fetched if an appropriate enzyme exists and was present".

The Biology of the Inner Light

Melatonin and pinoline, made by the pineal gland are regulated by the seasonal changes in light and darkness, linked to the sleep/wake cycle. Pinoline is made in the pineal from 5HT and hypothesized as the neurochemical trigger for dreaming. Lack of sleep for several nights is often linked to the onset of acute psychotic breakdown in which the person starts hallucinating or "dreaming while awake." This state of consciousness is common to the dream state, the psychedelic state, and the shamanic initiation experience. (Roney-Dougal)

Illumination has been described as being blinded by the manifestation of God's presence. This brightness has no relation to any visible light. Visionary experience, which has symbolic or religious content, may give way to this dazzling light, which is reported in eastern and western religions. It can confer a palpable glow to the person that is perceptible after the return to ordinary awareness.

Meditation modulates pineal activity, to create a standing wave through resonance effects that affects other brain centers with both chemical and electromagnetic coordination. Resonance can be induced in the pineal using electric, magnetic, or sound energy. Such harmonization resynchronizes both hemispheres of the brain. This results in a chain of synergetic activity resulting in the production and release of hallucinogenic compounds.

Sacred images are generated by the lower temporal area which also responds to ritual imagery, facilitated by prayer and meditation. Religious emotions originate from the middle temporal lobe and are linked to emotional aspects of religious experience, such as joy and awe. Yet neural correlates don't mean that these experiences exist "only" in the brain or are merely illusory; they are associated with distinct neural activity. There is no way to distinguish if the brain *causes* these experiences, or actually perceives spiritual reality. Visions of bright lights, portals, and spiritual icons correlate with DMT.

"Could it be that human beings have actually evolved specialized neural circuitry for the sole purpose of mediating religious experience?" Neurologist Ramachandran says so. "There may be certain neural pathways—neural structures in the temporal lobe and the limbic system—whose activity makes you more *prone* to religious belief."

If this is true, it is easy to see how much this mind-altering chemical could amplify all of the tendencies toward mystical apprehension originating in other parts of the brain.

The pineal contains high levels of the enzymes and building blocks for making DMT, and it may be secreted when inhibitory processes cease blocking its production. It may even produce other chemicals, such as beta-carbolines that magnify and prolong its effects.

Clear Light

Illumination has been described as being blinded by the manifestation of God's presence, which has no relation to visible light. Visionary experience, which has symbolic or religious content, gives way to this dazzling light, which is reported in eastern and western religions. Sacred Light is generated internally by DMT within the ventricles. Tendencies toward mystical apprehension originating in other parts of the brain are amplified. This universal Clear Light appears in all cultures with different names.

The mindbody is electronic, but it is rooted in the luminosity of its invisible ground. Living systems are very sensitive to tiny energy fields and resonance phenomena, both locally and at a distance. They allow the cells of the body to work together instantaneously and symphonically. All biological processes are a function of electromagnetic field interactions. EM fields are the connecting link between the world of form and resonant patterns. EM fields embody or store gestalts, patterns of information. Biochemical action and bioelectronic action meet at the quantum-junction.

We can return to Nature and our nature, collectively preparing a paradigm shift for a new shared reality and trajectory of physical, emotional, cognitive and spiritual coherence. The silent frictionless flow of living intelligence is beyond words and conceptual constructs. We are a process of recursive self-generation. This continuum, which is our groundstate or creative Source, is directly discoverable in the immediacy of the emergent embodied moment, in the living Light that generates our Being.

Blinded by the Light

Kabbalalists speak of this mystic Light during ecstatic entry into Pardes, the "orchard" of the Garden of Pomegranates, the self-luminous spheres of the Tree of Life. This metaphysical experience of the "Light of the Shekinah," the feminine aspect of the Divine, is associated with qabalistic ascent up the Middle Pillar. In this state the soul remains covered or adorned, and one cleaves to the Light, gazing at the awesome radiance of God (*Tzvi ha Shekinah*) in rapt mystic Union.

According to Kabbalist Idel, the grace of "sweet radiance" has erotic overtones. It also implies mystical death, separated from all concerns with the mundane world. The Divine Light attracts the light of the soul, "which is weak in relation to the Divine Light." The metaphor is one of magnetic attraction. The Kabbalists tried to reach the pre-fall state of the Primordial Man, to reenter the radiance of the Shekinah, a mystically erotic relationship with the Divine Presence which creates a reflective "glow."

Entrance of the philosopher or mystic into the Pardes affects only the human soul. But in the Theosophical paradigm it does have affects on the non-human realms, the system of divine powers, influencing the relationships between them. In the Theurgic paradigm there is also an influence on, or struggle with, the demonic realm, which seeks to hold the soul back from union.

In both cases, Pardes represents a danger zone, leading potentially to insanity or death, being overwhelming for most mortals. Premature entry to this realm has been likened to tearing a silk scarf from a rosebush, rather than gently removing it slowly (with regular meditation). It sounds like the wrathful visions of Buddhism and the intensely raw effects of unmediated DMT.

Yogatronics

Want to take an active role in your own spiritual life, a safe and easy mind trip? Would you like to glimpse some of the experiences outlined here? Or even just get the mental health benefits of deep relaxation and increased inner focus? Intimidated by the prospect of spending 15 to 20 years learning to meditate to attain life-enhancing benefits?

Haven't had a near-death experience and don't want one? Too busy to devote your life to alchemy, or spend endless years in transpersonal therapies, or too afraid to allow a "mad scientist" to zap your brain with EM frequencies, hook your brain up to a high-tech scanning machine, or inject you with psychedelic substances?

Modern technology offers an easy Do It Yourself, "passive" alternative. Anyone can employ a safe and easy technique that automatically puts you in the "zone." A form of "yogatronics" is available using a simple CD and headphones with input from subsonic frequencies. This audio technology creates a harmonization of the left and right hemispheres of the brain, and automatically drives the brain harmlessly into the Alpha or Theta brainwave range.

This resonance phenomenon, entrainment, is called the frequency-following response, or binaural beat technology. Entrainment is the process of synchronization, where vibrations of one object will cause another to oscillate at the same rate. It works by embedding two different tones in a stereo background. Continuous tones of subtly different frequencies (such as 100 and 108 cycles per second) are delivered to each ear independently via stereo headphones. The tones combine in a pulsing "wah wah" tone.

External rhythms can have a direct effect on the psychology and physiology of the listener. The brain effortlessly begins resonating at the same rate as the difference between the two tones, ideally in the 4-13 Hz. (Theta and Alpha) range for meditation. All you have to do is sit quietly and put on the headphones. The brain automatically responds to certain frequencies, behaving like a resonator.

You may not become immediately enlightened, but hemispheric synchronization helps with a whole host of problems stemming from abnormal hemispheric asymmetries. Problems, often resulting from stress or abuse in early life, include REM sleep problems, narcissism, addictive and self-defeating behaviors. Communication between hemispheres correlates with flashes of insight, wisdom and creativity.

Brain Synch

Our brain's two hemispheres are meant to work in concert with one another. Interactive hemispheric feedback is used to treat disorders such as post-traumatic stress disorder (PTSD), depression, ADD, addiction, obsessive-compulsive disorder, anxiety, and numerous other dysfunctions. Disorders of under-arousal include depression, attention-deficit disorder (ADD), chronic pain and insomnia. Overarousal includes anxiety disorders, problems getting to sleep, nightmares, ADHD, hypervigilance, impulsive behavior, anger/aggression, agitated depression, chronic nerve pain, and spasticity.

Because the brain is functionally "plastic" in nature, creating and exercising new neural pathways can retrain neural circuitry. In meditation, the halves of the brain become synchronized and exhibit nearly identical patterns of large, slow brainwaves. Rhythmic pulses can modulate collective neuronal synchrony. Then, both lobes automatically play in concert.

Rhythm regulates the entire spectrum of activation and arousal by kindling, or pulling more and more parts of the brain into the process. Disorders related to under- and over- arousal, including attentional and emotional problems, can be stabilized by self-organizing restructuring. Depressions, anxiety, worry, fear, and panic can be moderated. Stimulating neglected neural circuitry creates new pathways, improving equilibrium and long-term change, essentially "tuning" the nervous system.

There are many companies branding this self-regulation technology, both in "active" clinical neurofeedback programs, and as "passive" home programs. Among the oldest is the Monroe Institute <monroeinstitute.org>, which calls its trademarked method Hemi-Synch. Another program offered by Centerpointe Research Institute <centerpointe.com> is called Holosynch. BioPulse is another. Another variation uses light pulses from goggles to drive the process, and is marketed as Alpha-Stim.

Discussion

As of 2012, Barker and his colleagues at Cottonwood Research Foundation, Inc. upgraded their trials and protocols to determine the role of endogenous hallucinogens. They improved their measurement and low-level identification methodology a thousand-fold over previous attempts, using state-of-the-art liquid chromatography-mass spectrometry (LC/MS) equipment and claimed to meet their research goals.

They measured the three known endogenous hallucinogens and their major N-oxide metabolites in blood, urine, cerebrospinal fluid, ocular fluid and/or other tissues to fully assess the status of an endogenous hallucinogen pathway. They confirmed the structural identity of a major metabolite (the N-oxide) that has never before been monitored in any endogenous hallucinogen study in humans.

Their ayahuasca studies showed "a major metabolite of DMT, DMT-N-oxide (which retains the identifying structure of the parent substance), being excreted in the urine at levels 10-20 times greater than DMT itself after ayahuasca administration. Similarly, N-oxide levels in blood were four times greater than DMT. They note, "This is the first time this metabolite has been reported in humans following DMT administration by any route."

They are confident continuing studies will determine the normal role and function of these compounds in non-drug induced altered states, including dreams, psychosis, meditation, religious experience, childbirth, and near-death states. They cite Cozzi et al., stating that "the enzyme responsible for synthesis of the endogenous hallucinogens is present in pineal gland, brain, spinal cord and retinal tissues of primates and appears to be an inducible enzyme, an enzyme that responds to specific signals. Therefore, clearly establishing the role of endogenous tryptamine hallucinogens in various states of consciousness will provide tremendous insight into their origin, and may lead to more reliable means of working with and studying their utility".

Conclusions

Are there things we should not know? We are innately geared to crave ecstasy, "escape reality," and seek extraordinary or novel experiences on our way to wisdom. The history of mankind recounts the stages of that journey. Religions, mystery schools, and mysticism arose from accounts of spontaneous spiritual experiences. In shamanism, our ancestors sought them in an instinctual or animalistic way. In art, myth and ritual we sought them in a human, if narcissistic and self-expressive reactionary way.

Curiously, DMT is ubiquitous in the biosphere, found everywhere from a variety of botanicals to mammals: It has been documented in rat brains at birth. Not only is it found in seaweed, flowers, vines, acacia tree (*Sant*), toadskins, *Desmanthus illinoensis* and *Mimosa hostilis*, A. columbrina, lawn grass, etc., but also in our brains and spinal columns.

*"It is only in Western society that the potential shaman, with all of their psychic gifts, is ignored and treated as sick. All other human societies have honored their prophets, psychics, seers and shamans. We need to learn to recognize the potential shaman in our midst and re-learn what is required to ground them, teach them and train them so that their creative and psychic abilities can be a gift, not a curse, and can be used for their and our benefit." (*Roney-Dougal)

In creativity and meditation we seek in a fully conscious way, willfully cooperating and facilitating the process not only of connecting with the divine, but experiencing ourselves in the process of "becoming" divine or being sacred. The ego no longer perceives itself as a separate expression of consciousness, but reconnects in a stabilized, not sporadic way. Our metaprogram is the same essence as All, infused with Light.

References

Abraham, Ralph, 2006, Recollections of the impact of the psychedelic revolution on the history of mathematics and my personal story. http://www.ralph-abraham.org/articles/MS%23124.Maps/maps2.pdf

Baconnier, Lang et al., Some thoughts on the paper: Calcite Microcrystals in the Pineal Gland of the Human Brain, First Physical and Chemical Studies, Bioelectromagnetics 23:488495 (2002). Biolectromagnetic Crystals in the Pineal May Be Resonant Piezoelectrics.

Barker, S.A., Ethan H. McIlhenny, Rick Strassman, 2012, A Critical Review of Reports of Endogenous Psychedelic N, N-Dimethyltryptamines in Humans: 1955-2010, Drug Testing and Analysis, in press (invited paper in Special Issue on Psychedelic Drugs).

Bosman, Ananda, Pineal Power, http://www.akasha.de/~aton/PINEALpower.html

Callaway, J.C. (1994). Pinoline and Other Tryptamine Derivatives: Formations and functions. PhD Dissertation, Dept. Pharmacol. & Toxicol, Univ. Kuopio, Finland

Chia, Mantak, (2002), Dark Room Enlightenment, Universal Tao Center, Thailand, http://www.scribd.com/doc/4474044/Dark-Room-Enlightenment-Mantak-Chia

Collected Abstracts, Scientific Evidence of Psychedelic Body Fluids, http://deoxy.org/annex/daytripr.htm#5

Cozzi, N.V., T. A. Mavlyutov, M. A. Thompson, A. E. Ruoho. Indolethylamine N-methyltransferase expression in primate nervous tissue. Soc. Neurosci. Abs. 2011, 37, 840.19 (2011)

Graves, Robert, (1948) The White Goddess: A Historical Grammar of Poetic Myth.
Journal of Pineal Research http://www.blackwellpublishing.com/journal.asp?ref=0742-3098

Grof, Stanislav, (1988), The Adventure Of Self-Discovery: Dimensions of Consciousness And New Perspectives In Psychotherapy, State Univ of New York Pr.

Korkmaz, Ahmet and Russel J. Reiter, (2007), Epigenetic regulation: a new research area for melatonin?, Journal of Pineal ResearchVolume 44, Issue 1, Article first published online: 26 OCT 2007.

McIlhenny, Ethan H., Jordi Riba, Manel J. Barbanoj, Rick Strassman, and S.A. Barker, 2012, Methodology for the Determination of Ayahuasca's Major Constituents and Their Metabolites in Blood, Journal of Biomedical Chromatography, Published online 2011, Jun 28. doi: 10.1002/bmc.1657.

McIlhenny, Ethan H., Jordi Riba, Manel J. Barbanoj, Rick Strassman, and S.A. Barker, 2011, Methodology for and the Determination of the Major Constituents and Metabolites of the Amazonian Botanical Medicine Ayahuasca in Human Urine, J. Biomed. Chromatogr. 25, 970-984.

McKenna, Terence; Food of the Gods; New York: Bantam Books, 1993.

Meyer, Peter (1993), "Apparent communication with discarnate entities induce by DMT", Psychedelic Monographs & Essays, Vol. 6, Thom Lyttle, Ed., PM & E Publishing Group, Boynton Beach, Florida.

Miller, Iona; "Chaos as the Universal Solvent: Re-creational ego death in psychedelic consciousness"; Psychedelics ReImagined, Thom Lyttle, Ed. New York: Autonomedia, 1999.

Miller, Iona (1994) Becoming the Vine: An Anecdotal Account of an Ayahuasca Initiation

Miller, Iona (2001), Neurotheology101: Technoshamanism and Our Innate Capacity for Spiritual or Mystical Experience, Institute for Consciousness Science & Technology, Wilderville, Oregon.

Miller, Iona (2006), How the Brain Creates God: The Emerging Science of Neurotheology, Chaosophy, Asklepia Pub. http://neurotheology.50megs.com

Pickover, Cliff (2006) http://sprott.physics.wisc.edu/Pickover/pc/dmt.html

Radha, Soami Sivananda (1990) The Divine Light Invocation; Timeless Books, Spokane, Washington.

Riba, Jordi, Ethan H. McIlhenny, Marta Valle, José Carlos Bouso S. A. Barker, 2012, Metabolism and disposition of N,N-dimethyltryptamine and harmala alkaloids after oral administration of ayahuasca, Drug Testing and Analysis, submitted (invited paper in Special Issue on Psychedelic Drugs).

Roney-Dougal, Serena, Walking Between the Worlds: Links Between Psi, Psychedelics, Shamanism, and Psychosis.

Sabom, M. B. (1982). Recollections of death: a medical investigation. Harper and Row, New York.

Santoro, R.., Marani, G Blandino, P Muti and S Strano, (2012) Melatonin triggers p53Ser phosphorylation and prevents DNA damage accumulation, Oncogene 31, 2931-2942 (14 June 2012) | doi:10.1038/onc.2011.469

Strassman, Rick (1990), "The Pineal Gland", Psychedelic Monographs & Essays, Vol. 5, Thom Lyttle, Ed., PM & E Publishing Group, Boynton Beach, Florida

Strassman, Rick (2001). DMT: The Spirit Molecule. Rochester, Vermont: Park St. Press. http://www.rickstrassman.com

Szára, Stephen, The Social Chemistry of Discovery: The DMT Story. (1989)

Made in the USA
Middletown, DE
20 September 2022